Florida
1850 Agricultural Census

Transcribed and Compiled by
Linda L. Green

WILLOW BEND BOOKS
2007

WILLOW BEND BOOKS
AN IMPRINT OF HERITAGE BOOKS, INC.

Books, CDs, and more—Worldwide

For our listing of thousands of titles see our website
at
www.HeritageBooks.com

Published 2007 by
HERITAGE BOOKS, INC.
Publishing Division
65 East Main Street
Westminster, Maryland 21157-5026

International Standard Book Number: 978-0-7884-4306-0

Introduction

This census names only the head of the household. Often times when an individual was missed on the regular U. S. Census, they would appear on this agricultural census. So you might try checking this census for your missing relatives. Unfortunately, many of the Agricultural Census records have not survived. But, they do yield unique information about how people lived. There are 48 columns of information. I chose to transcribe only six of the columns. The six are: Name of the Owner, Improved Acreage, Unimproved Acreage, Cash Value of the Farm, Value of Farm Implements and Machinery, and Value of Livestock. Below is a list of other types of information available on this census.

Linda L. Green
13950 Ruler Court
Woodbridge, VA 22193

Other Data Columns

Column/Title

6. Horses
7. Asses and Mules
8. Milch Cows
9. Working Oxen
10. Other Cattle
11. Sheep
12. Swine
14. Wheat, bushels of
15. Rye, bushels of
16. Indian Corn, bushels of
17. Oats, bushels of
18. Rice, lbs of
19. Tobacco, lbs of
20. Ginned cotton, bales of 400 lbs each
21. Wood, lbs of
22. Peas and beans, bushels of
23. Irish potatoes, bushels of
24. Sweet potatoes, bushels of
25. Barley, bushels of
26. Buckwheat, bushels of
27. Value of Orchard products in dollars
28. Wine, gallons of
29. Value of Products of Market Gardens
30. Butter, lbs of
31. Cheese, lbs of
32. Hay, tons of
33. Clover seed, bushels of
34. Other grass seeds, bushels of
35. Hops, lbs of
36. Dew Rotten Hemp, tons of
37. Water Rotted Hemp, tons of
38. Other Prepared Hemp
39. Flax, lbs of
40. Flaxseed, bushels of
41. Silk cocoons, lbs of
42. Maple sugar, lbs of
43. Cane Sugar, hunds of 1,000 lbs
44. Molasses, gallons of
45. Beeswax, lbs of
46. Honey, lbs of
47. Value of Home Made Manufactures
48. Value of Animals Slaughtered

Table of Contents

Alachua County, Florida
1850 Agricultural Census

The University of North Carolina at Chapel Hill filmed the 1850 agricultural census for Alachua County from originals at the Florida State University under a grant from the National Science Foundation in 1963.

Columns 1, 2, 3, 4, 5, and 13 represent the following information on the census:
1. Name of Owner, Agent or Manager of Farm
2. Acres of Improved Land
3. Acres of Unimproved Land
4. Cash Value of the Farm
5. Value of Farming Implements and Machinery
13. Value of Livestock

E. R. Prevatt, -, -, -, -, 300
John M. Hendry, -, -, -, -, 150
Horace Mcory, 50, 833, 1500, 150, 3090
William B. Hart, 35, 945, 1000, 100, 1400
Cathrine Payne, 14, 90, 600, 50, 660
John S. Livingston, 15, 90, 1000, 300, 965
James A. Turner, 15, 74, 500, 60, 1700
John H. May, 9, -, 50, 20, 233
Ransom Harris, 10, -, 150, -, 150
John B. Frye, -, -, -, -, 25
R. S. Stoughton, 14, 90, 600, 200, 695
Joseph Lancaster, 23, -, 150, -, 400
John H. Zanadski, 6, -, 800, 350, 700
Michael McCarrol, 28, -, 200, 30, 790
Joshua Stafford, 8, -, 151, -, 1600
William S. Turner, 150, 1300, 2000, 100, 1600
Benjamin Mills, 60, 350, 520, 150, 1130
James Jones, 10, -, 100, 12, 175
Albert G. Roberts, 20, 129, 350, 25, 150
William H. Carpenter, 16, 48, 1200, 1100, 515

David Mizell, 50, 1505, 2500, 100, 1410
James Tompkins, 10, 46, 200, 50, 700
John A. Dean, 8, 150, 150, 25, 765
Cornelius Barker, 25, 125, 250, 50, 430
Edward Lane, 8, -, 150, 10, 40
James T. Thomas, 20, 130, 300, 60, 3440
Daniel Conyers, 25, 135, 500, 75, 250
James B. Hutson, 5, -, 300, 10, 187
Samuel Hudson, 30, 120, 600, 50, 540
Henry Sparkman, 25, 55, 250, 25, 130
William Williamson, 14, 146, 225, -, 310
Daniel Morrison, 80, 280, 2000, 800, 735
McKeen W. Carlton, 20, -, 150, 15, 262
Counsel Stokes, 30, 129, 500, 40, 336
John Carrol, 30, 120, 400, 100, 480
Arthur Pinnes, 4, -, -, -, 170
Israel Sherhoull, -, -, -, -, 225
Enoch Collins, 40, -, 200, 100, 1090
Joseph Mizell, 25, 135, 400, 100, 1190

Burnell Green, 22, 138, 350, 30, 125
Daniel Sparkman, 3, -, 50, 5, 12
John Chesser, 25, 615, 300, 50, 450
James A. Clark, 20, 620, 150, 25, 19
Peter Sparkman, 65, 2575, 1000, 300, 1366
Arnold Thigpin, 12, 28, 100, 25, 130
William Sparkman, 10, -, 100, 30, 100
Bryant Glisson, 8, -, 100, 10, 110
Joshua Platt, 10, -, 200, 100, 335
Jesse Hagan, 14, -, 200, 50, 415
Michael Peterson, 45, -, 400, 40, 85
Louis Blackshear, 60, 30, 1500, 100, 357
Benjamin Tyson, -, -, -, 25, 270
Joseph Thomas, 12, -, 300, 50, 675
Henry Sylivant, 4, -, 100, 10, 220
George Durance, 9, -, 160, -, 250
John McCoven, -, -, 150, 10, 275
Andrew Weells, 10, -, 75, 15, 120
W. C. Allen, 35, -, 300, 10, 282
John N. Prescott, 5, -, 75, 8, 320
John M. Geiger, 100, 20, 1200, 300, 2335
V. R Prevatt, 26, 16, 500, 40, 930
Levi J. Johns, 10, -, 300, 45, 425
Wade H. Sparkman, 60, -, 600, 250, 120
Joseph J. B. Holder, 4, -, 150, 20, 494
Stephen Sparkman, 30, -, 400, 100, 1258
A. J. James, 5, -, 125, 25, 262
Ann Munroe, 80, 80, 500, -, 80
Thomas W. Green, 36, -, 250, 75, 196
Nathaniel Jones (James), 60, -, 700, 55, 635
W. S. Bernnett, 35, -, 100, 40, 340
Samuel Thomas, 15, -, 160, 30, 190
Samuel Thomas, 10,-, -, 40, 265
Thomas Tillis, 10, -, 150, 20, 375
Toplin Tillis, 12, -, 100, 10, 75
Elizabeth Moory (Mcory), 20, -, 200, 15, 182

Joseph Tullis, 4,-, 100, 35, 296
Asa Clark, 150, 200, 3000, 500, 1400
Stephen Dampier, 8, -, 75, 15, 100
Isaac Johns, 8, -, 150, 50, 120
Thomas B. Holder, 20, -, 250, 45, 1250
John G. Brookes, 6, -, 100, 75, 138
James H. Smith, 56, 64, 1200, 50, 530
George A. Hires, 20, 20, 500, 20, 290
Mitchel Kirtland, 20, -, 150, 50, 214
Abraham Mott, 40, 160, 1500, 100, 820
James Bevil, 25, -, 200, 50, 240
James Moore, -, -, 40, 20, 160
Henry Moore, 16, 64, 200, 10, 170
Luther Randol, 5, -, 50, 4, 12
Joseph G. Bell, 70, 40, 2000, 70, 790
William F. Smith, 42, 40, 700, 120, 28
Paul B. Colson, 16, 64, 300, 110, 456
Arthur Floyd, 80, 120, 800, 60, 346
Hope Mott, 10, -, 100, 5, 55
William Dell, 40, 120, 1000, 500, 1635
Samuel Russell, -, -, -, -, 315
A. E. Geiger, 80, 200, 500, 125, 361
Elija Barrow, 35, 45, 500, 100, 385
Samuel W. Burnett, -, -, -, -, 245
George L. Brown, 5, -, -, 300, 735
Thomas Paisley, 40, -, 100, 30, 125
Thomas J. Prevatt, 200, 640, 3000, 200, 5040
William F. Rowan, -, -, -, -, 117
George W. Galpin, -, -, -, 50, 200
Lemuel Wilson, 30, -, 700, 25, 550
William C. Mott, 15, -, 200, 85, 180
James Burnett, 12, 28, 250, 40, 380
Sarah Colson, 12, 68, 300, 30, 224
James Cason, 40, 80, 1000, 125, 878
Ranson Cason, 75, 45, 1000, 100, 860

John C. Richard, 120, 320, 3000, 670, 1893

Elisha Carter, 10, -, 40, 5, 10

Clemma Douglass, 16, -, 150, 150, 630

Thomas Harvell, 17, -, 100, 5, 205

John Cason, 25, -, 130, 50, 622

Samuel R. Piles, 200, 600, 5000, 429, 1610

Walter Button, 30, -, 150, 50, 280

Thomas W. Piles, 130, 190, 4000, 292, 1037

William Colson, 300, 340, 2500, 200, 1050

Simeon Dell, 120, 520, 2000, 150, 955

Louisa J. Tyson, 30, 25, 250, 75, 430

Martin Wimberly, 20, -, 300, 200, 508

Jacob Strobel, 40, -, 250, 100, 279

Mary Wymer, 30, -, 700, 50, 70

F. A. Underwood, 50, 65, 1000, 200, 1237

Oliver Bryant, 16, -, 200, 150, 432

Wade Bryant, 15, -, 30, 5, 112

George Sharp, 16, 164, 500, 50, 292

Daniel Strobel, 22, -, 200, 10, 146

Charles F. Fitchet, 35, 45, 500, 50, 332

David Ridout, -, -, -, -, -

William J. Morgan, 15, -, 150, 40, 20

John B. Stanley, 250, 600, 3000, 1000, 3756

Zilpha Stanley, 400, 1000, 5000, 2000, 4518

Isaiah Floyd, 8, 32, 150, 5, 830

Thomas W. Smith, 13, -, 100, 10, 232

George Watson, 13, -, 200, 100, 283

Presley Thomas 8, -, 35, 15, 116

Moses Slaughter, 40, -, 150, 125, 1005

Rachael J. Townsend, 20, -, 100, 75, 360

William Cannon, 14, -, 250, 45, 565

Noah P. Sugs, 40, -, 300, 30, 366

Robert Bevan, 15, 265, 800, 50, 790

Amy Hagan, 28, -, 150, 50, 378

William J. Hart, 16, -, 200, 6, 100

John Rogers, 20, -, 300, 10, 394

Maria Sombray, 4, -, 200, 5, 38

Bernett M. Dell, 400, 1892, 20000, 700, 2460

Calvin Bryan, 60, 220, 1400, 1000, 623

James Hague, 30, 178, 2500, 200, 650

John R. Hague, 23, 17, 150, 10, 420

George W. Sanchez, 125, 1875, 6000, 400, 820

Benjamin O. White, 40, 40, 200, 150, 235

Gilford Sykes, 24, -, 700, -, 253

Solomon Warren, 75, 568, 300, 200, 1090

Charles L. Wilson, 100, 220, 1000, 150, 996

Jonathan Turner, 50, 390, 1200, 60, 664

Eliza Tucker, 27, 75, 300, 40, 445

Samuel Geiger, 165, 315, 2000, 300, 2054

Mathew Chesser, 40, 600, 1000, 200, 1417

Robert Burton, 7, -, 100, 10, 73

John H. Malphues, 15, 145, 400, 40, 227

Burel Stokes, 12, -, 180, 8, 203

William Malphues, 20, 300, 400, 50, 675

David Syms, 20, 640, 300, 100, 640

William Sapp, 18, -, 50, 10, 126

Josiah Secenger, 18, 143, 350, 100, 524

Andrew Robb, 6, -, 50, 100, 570

Elizabeth Duskin, 26, -, 150, 110, 76

John Dixon, 14, -, 150, 24, 182

John Wiggins, 38, -, 400, 100, 2032

James S. Chesser, 6, -, 25, -, 25

John McKinley, 20, 620, 175, 10, 128

James G. Cameron, 160, 1760, 3000, 1000, 1320
James B. Colding, 30, 610, 50, 50, 353
Thomas Colson, 15, -, 100, 6, 138
John Prior, 6, -, 100, 35, 280
Hope H. Colson, 30, 550, 250, 60, 4680
S. B. Osteen, 65, 474, 900, 10, 275
Cornelius Rains, 100, 430, 1000, 200, 4900
Henry B. Turner, 50, 148, 1000, 58, 580
Jeptha Knight, 30, -, 150, 50, 1435
Samuel Chesser, 30, -, 40, 40, 300
John Little, 23, 617, 500, 40, 212
James S. Gibson, 14, -, 20, 10, 98
D. B. Williamson, 15, 25, 250, 25, 364
George Walker, -, -, 25, 5, 46
John Light, -, -, 25, 5, 138
W. D. Sistrunk, -, -, 15, 2, 120
James A. Osteen, -, -, 15, 5, 144
James Edwards, 12, -, 288, 25, 110
John A. Seaton, 10, -, 200, 30, 220
Richard H. Parker, 70, 10, 500, 130, 541
William D. Clark, 20, 100, 1000, 100, 550
Cornelius Johns, 12, -, 400, 5, 360
John W. Hurst, 12, -, 100, 40, 500
John H. Stevens, -, -, 250, 20, 270
Jacob Halbrook, 20, 20, 200, 30, 540
Philip Dell, 225, 95, 3500, 500, 1175
Clates Shuhouse (Sherhouse), -, -, -, -, 115
Sarah Hooker, 10, 630, 500, 25, 558
Abraham Guthrey, 10, -, 55, 2, 60
Reason L. Terrill, 10, -, 60, 20, 45
John Sparkman, 30, -, 500, 60, 1453
Godliff Sherhouse, 15, -, 200, 100, 550
James Fennell, 10, -, 300, 25, 340
Jehu Sparkman, 12, -, 200, 35, 300
James M. Sparkman, 40, -, 600, 150, 1100

Silas Weeks, 60, 10, 1000, 100, 1030
Garrey Ford, 10, -, 50, 10, 100
Thomas Walker, 6, - 75, 10, 105
John Ketler, 50, 29, 1000, 120, 375
Franklin Desher, 10, -, 100, 30, 260
Richard Howard, 12, -, 150, 40, 330
Ezekiel Parrish, -, -, -, 25, 48
Riley Moore, 22, -, 75, -, 240
Wiley Hicks, 25, -, 200, 30, 165
Henry Snowden, 10, 630, 150, 20, 115
James Crosby, - -, -, 40, 275
William Adams, 20, 600, 250, 150, 800
James M. Flinn, 6, 634, 150, 50, 655
John A. Kenady, 50, -, 200, 200, 840
David Gillett, 56, 624, 500, 175, 2960
Jacob Link, 20, 620, 400, 125, 430
Joshia Pery, 30, -, 100, 57, 140
Henry Sweeny, 45, 435, 600, 125, 735
Jonathan Thigpin, 5, -, 40, 10, 315
James F. Secrest, 40, 1654, 650, 100, 330
John R. Zatronner, 15, 625, 1500, 100, 1838
M. S. Pery, 170, 2206, 1500, 235, 1575
Levi Sparkman, 5, -, 20, 25, 105
Charles J. McMims, 10, 613, 100, 30, 355
John Strickland, 15, 145, 150, 5, 70
William D. Eubank, 50, 605, 1500, 200, 1020
Samuel Saunders, 12, 316, 200, 50, 200
Reding Tuton, 8, 160, 100, 25, 238
John R. Saunders, 30, 900, 300, 100, 420
Mary Link, 7, -, 50, 10, 4910
Moses Aldrich, 7, -, 50, -, 200
Moses Ramsey, 120, 620, 1100, 300, 1466
Henry Bouknight, 35, 1245, 3000, 225, 550

James Tyson, 12, -, 200, 40, 655
John F. McDonald, 75, 520, 1500, 200, 1460
Daniel Skipper, 3, -, 40, 5, 260
Joseph A. Evritt, 37, 80, 1000, 300, 1880
Lawson Clark, 15, -, 120, 30, 596
Samuel B. Colding, 30, 617, 1000, 50, 1230
Stephen McKinney, 6, -, 25, -, 50

James McKinney, 16, -, 100, -, 262
Hardy Harvard, 20, 300, 600, 150, 1109
A. B. Fussell, 24, -, 100, 40, 260
Simon Hodges, 20, 620, 240, 30, 667
Joel B. Smith, 20, 725, 700, 20, 416
Stephen Fagan, 50, 30, 700, 200, 955
Miles Shepherd, 70, -, 100, 30, 90
Nathaniel Nobles, -, -, -, -, 30

Benton County, Florida
1850 Agricultural Census

The University of North Carolina at Chapel Hill filmed the 1850 agricultural census for Benton County from originals at the Florida State University under a grant from the National Science Foundation in 1963.

Columns 1, 2, 3, 4, 5, and 13 represent the following information on the census:
1. Name of Owner, Agent or Manager of Farm
2. Acres of Improved Land
3. Acres of Unimproved Land
4. Cash Value of the Farm
5. Value of Farming Implements and Machinery
13. Value of Livestock

Henry Hope, 35, 125, 1200, 50, 2738
William K. Grey, 35, 125, 400, 100, 632
Light Townsend, 20, 140, 550, 20, 476
John W. Crichton, 100, 60, 1500, 106, 1036
William W. Tucker, 125, 195, 4000, 50, 1120
Edward M. Harville, 20, -, 100, 12, 260
William Hope, 140, 420, 3360, 150, 4143
Richard R. Crown, 12, 40, 300, 55, 512
Rufus Boyer(Hoyer), 6, 434, 1740, 5, 451
David Hope, 20, -, 125, 45, 860
John Boyer, 16, 144, 1000, 20, 400
James A. Boyer, 30, 130, 1200, 50, 620
John J. Selph, 12, 148, 400, 15, 335
John J. Selph, 5, 755, 450, 55, 460
Ezekiel S. Selph, 8, 152, 350, 8, 502
Jason McKinney, 7, 153, 500, 30, 135
Jack L. Andrews, 33,-, 300, 5,-
James M. Clausen, 20, 18, 800, 63, 445
Daniel Simmons, 70, 50, 1000, 200, 650

Nathan Baget, 9, -, 75, 4, 259
John B. Allen, 50, 430, 1500, 200, 380
Cornelius Seals, 80, 240, 1500, 100, 870
Jackson Andrews, 20, -, 50, 10, 75
Edward Baget, 4,-, 100, 5, 96
Beril M. Pearson, 750, 210, 2350, 1340, 1050
Joel S. Lockhart, 32, 128, 500, 20, 100
G. W. Andrews, 50, 110, 1500, 110, 358
Watson W. Legett, 25, -, 200, 20, 220
Albert Clark, 40, 120, 1500, 140, 1330
William H. Main, 10, 150, 600, 5, 150
Robert G. Oswan, 19, 140, 600, 45, 150
William Prescott, 12, 309, 1600, 35, 115
James Burke, 8, 152, 600, 5, 1800
Joseph Hall, 35, 215, 1500, 300, 1020
Nancy P. Harville, 125, 515, 6400, 175, 359
David L. Yuler, 100, 540, 3500, 300, 500
John Morris, 6, -, 150, 25, 190

James Baker, 30, 180, 800, 40, 430

Abraham Hay, 5, 120, 120, 55, 120

Elias James Knight, 30, 130, 500, 180, 930

James Haymons, 15, 145, 500, 365, 171

Albert J. Alexander, 50, 200, 200, 200, 375

Columbus R. Alexander, 6, 34, 200, 10, 290

David B. Turner, 10, -, 150, 50, 450

Peter W. Law, 60, 100, 3000, 100, 1330

Dewey G. Wase (Ware), 80, 200, 3000, 160, 925

Malcolm Peterson, 20, 180, 1000, 80, 655

William Taylor, 22, 298, 1000, 50, 450

John _. Taylor, 100, 1163, 3500, 25, 600

Jeremiah Dodson, 25, 135, 1200, 25, 444

Thomas C. Ellis, 50, 180, 1500, 100, 1005

Joseph Brownlow, 3, -, 100, 5, 320

John A. Suton, 130, -, 500, 30, 305

John Bassett, 20, 140, 1100, 100, 430

Isaac Garrison, 40, 280, 1100, 50, 3520

Nancy Wiggins, 20, 140, 500, 30, 530

James McMullins, 25, -, 300, 85, 236

Robert D. Bradley, 50, 30, 900, 88, 850

Charles Goodrich, 20, -, 200, 90, 175

N. K. Sparkman, 19, 141, 400, 35, 1095

Shadrack Sutton, 3, -, 100, 30, 450

Willis Smith, 8, -, 158, 45, 940

Thomas Tucker, 14, -, 125, 45, 1900

George Dyke, 20, 140, 800, 40, 220

Lyburn Hersey, 20, -, 250, 35, 200

Jacob Wells, 15, 145, 600, 66, 228

Mathew Jones, 15, 25, 260, 20, 280

John G. Tyner, 15, 145, 600, 100, 320

William Hendricks, 20, 300, 2000, 100, 2000

Mary Sylvester, 20, -, 300, 30, 959

John Lanier, 15, -, 200, 100, 4370

Isaac Lanier, 8, -, 100, 50, 3190

James Lanier, 15, -, 100, 100, 4340

Nathl. M. Moody, 25, 160, 200, 25, 960

Enoch E. Mizell, 12,-, 150, 45, 1365

Mills Holloman, 8, -, 100, 45, 858

Morgan Mizell, 11, 149, 649, 60, 896

Joshua Mizell, 11, -, 130, 30, 618

John Townsend, 12, 148, 560, 100, 1550

John Tucker, 20, 60, 300, 40, 520

Andrew A. Crown, 10, 160, 880, 20, 60

Jane (James) M. Bales, 30, 20, 300, 60, 450

Calhoun County, Florida
1850 Agricultural Census

The University of North Carolina at Chapel Hill filmed the 1850 agricultural census for Calhoun County from originals at the Florida State University under a grant from the National Science Foundation in 1963.

Columns 1, 2, 3, 4, 5, and 13 represent the following information on the census:
1. Name of Owner, Agent or Manager of Farm
2. Acres of Improved Land
3. Acres of Unimproved Land
4. Cash Value of the Farm
5. Value of Farming Implements and Machinery
13. Value of Livestock

Ann Kern, -, -, 150, 30, 135
William B. Goodrum, -, 27, 400, 40, 1299
William M. Christian, -, -, 200, 75, 294
William Ayrs, -, -, 200, 30, 779
Robert A. Farley, -, 40, 700, 50, 160
Jonathan Williams, -, 40, 100, 10, 193
William P. Simmons, 40, -, 1600, 40, 415
Stephen Richards, 80, 620, 5000, 150, 1150
Rubin Gay, 65, 135 3600, 51, 814
John Strawn, -, -, 200, 30, 138
James Durdan, -, -, 200, 20, 226
John Durdan, -, -, 150, 10, 629
Jason Gregory, 150, 1351, 65000, 300, 700
Sarah Caraway, 5, 35, 500, 50, 458
A. B. Caraway, 130, 140, 1000, 150, 664
Mary Ann Yon, 30, 426, 500, 50, 571
Jea Atkins, 70, 290, 1000, 50, 446
William B. Sanson, 13, 80, 200, 100, 841
Stephen Racrek (Roark), -, -, 150, 5, 34
John T. Carruth, 12, 8, 800, 100, 250
Jesse F. Walker, 8, 52, 400, 26, 411

Archabald Kelly, 20, 100, 600, 50, 581
Ann M. Lott, 200, 1000, 8000, 250, 935
John Griffin, -, -, 100, 50, 202
Jehu (John) Richards, 50, 150, 800, 150, 790
Solomon Whichard, -, -, 100, 150, 135
Harison Burgess, 100, -, 500, 30, 670
William Messer, -, -, 150, 20, 191
Jacob Scott, -, -, 100, 50, 182
Jacob Scott, -, -, 250, 20, 382
John G. Richards, 30, 130, 960, 400, 531
Daniel T. Richards, 35, 5, 800, 31, 340
George W. Underwood, -, -, 100, 75, 675
William W. Ramsey, -, -, 100, 40, 145
William Castlebury, -, -, 100, 35, 236
John Wood, -, -, 10, -, 120
Horace Ely, 40, 280, 1250, 50, 371
Daniel Smith, 50, 31, 800, 100, 1000
William Calahan, -, -, 100, 200, 185
Alford Edwards, -, -, 100, 20, 95
William B. Wynn, 300, 550, 10000, 375, 1980
Elias Branch, 6, 51, 500, 60, 740

Herod Dales, -, -, 500, 50, 645
William McKinny, -, -, 300, 10, 153
John Clarke, 180, 277, 4000, 570, 638
John Ott, 12, 133, 1000, 20, 235
John Gasque, 10, 70, 400, 31, 146
Henry Avant, 25, 255, 2800, 30, 379
Samuel Achoan, -, -, 400, 20, 175
Terrel H. Yon, 40, 900, 6000, 20, 405
Almerine J. Wood, 80, 320, 4000, 155, 1150
John Spears, 12, 144, 1560, 50, 542
Joseph Davis, 20, 140, 1200, 10, 580

Daniel Meeks, -, -, 230, 50, 174
Britton Barkley, 60, 100, 2500, 200, 550
Needham McKinney, -, -, 400, 51, 262
William H. Burnett, -, -, 960, 35, 380
William Walia (Waha), 30, 327, 3500, 71, 532
Joel Porter, -, -, 1500, 50, 1675
Samuel W. Davis, 22, 290, 100, 100, 325
James Nall, 22, 238, 300, 200, 1640
Alexander Ward, 20, 100, 500, 30, 195

Columbia County, Florida
1850 Agricultural Census

The University of North Carolina at Chapel Hill filmed the 1850 agricultural census for Columbia County from originals at the Florida State University under a grant from the National Science Foundation in 1963.

Columns 1, 2, 3, 4, 5, and 13 represent the following information on the census:
1. Name of Owner, Agent or Manager of Farm
2. Acres of Improved Land
3. Acres of Unimproved Land
4. Cash Value of the Farm
5. Value of Farming Implements and Machinery
13. Value of Livestock

John C. Pelote, 100, 60, 800, 200, 450

Asa A. Stewart, 50, 100, 50, 50, 800

John W. Lowe, 75, 85, 2000, 500, 800

Douglass O'Neil, 15, 25, 800, 30, 304

William H. Hutchins, 60, 20, 200, 300, 113

George W. Donaldson, 12, 300, 390, 10, 250

William Light, 20, 20, 200, 15, 350

Britton Knight, 40, 200, 500, 50, 200

Andrew Y. Allen, 175, 120, 1500, 125, 875

William H. T. Robarts, 50, 20, 800, 70, 700

Cannon Tyre, 30, 4, 250, 40, 467

William A. Gooldsby, 45, 35, 2100, 40, 354

Louis Daughtrey, 45, 35, 700, 370, 620

Stering Scarborough, 30, 50, 1000, 30, 295

Joseph J. Knight, 50, 110, 1500, 15, 200

Laurence K. Mickles (Mickler), 33, 127, 800, 40, 350

Thomas Bryant, 25, 80, 400, 80, 810

Elias E. Johnston, 15, 1, 150, 20, 424

William H. Keene, 90, -, 1000, 50, 660

Levi Wright, 40, 40, 500, 20, 1000

Stephen McGoven, 18, -, 170, 15, 148

Littleton Smith, 12, 50, 150, 15, 423

Abram Rivers, 120, 50, 1500, 40, 462

James S. Jones, 125, 118, 1400, 125, 592

Elisha Parker, 15, 25, 250, 50, 407

John W. Powers, 110, 200, 1000, 150, 782

William Lee, 8, 30, 300, 30, 230

Shubal Barnes, 100, 180, 5000, 200, 1068

Oliver Waldron, 35, 125, 200, 40, 698

John S. Goodbread, 50, 75, 700, 50, 700

Peter Cannon, 13, 27, 200, 15, 140

John Bryant, 20, 20, 200, 30, 510

Thomas H. Gooldsby, 12, -, 50, 10, 132

Henry Mattair, 25, 15, 450, 120, 668

William B. Ross, 300, 700, 4000, 1000, 1170

Martin Hancock, 27, -, 100, 100, 1175

John Roberts, 170, 310, 6000, 500, 2550

Zachary R. Roberts, 65, 135, 700, 50, 610

Reuben Cottle, 35, 20, 400, 10, 790

William Peirce, 45, 75, 600, 75, 687

James C. Peirce, 80, 90, 500, 75, 1668

William Jones, 25, -, 300, 25, 1420

Arthur Roberts, 75, 85, 2000, 200, 463

Langley Bryant Sr., 21, 59, 400, 10, 450

Hezekiah Osteen, 20, 60, 300, 30, 500

Wright Douglass, 20, 140, 700, 30, 292

Henry Boyd, 25, 15, 500, 125, 560

John Niblack, 30, 130, 400, 20, 490

Stephen Sparkman, 80, 240, 800, 300, 2675

John Markhan, 17, 63, 250, 9, 710

William Markhan, 45, 35, 500, 100, 1380

George W. Robarts, 40, 80, 400, 40, 435

Langley Bryant Jr., 33, 60, 350, 40, 1140

Shaderick Sapp, 25, -, 200, 50, 660

Samuel S. Wester, 75, 300, 500, 75, 495

Durham Hancock, 50, 150, 550, 50, 575

Liberty F. Raulerson, 35, 45, 500, 25, 500

Anderson Gillett, 40, 50, 600, 60, 1388

Andrew J. Keene, 14, 26, 200, 40, 440

Ansel Walker, 11, 69, 250, 50, 765

Louis Clark, 20, 40, 180, 30, 310

Samuel Niblack, 23, 57, 250, 50, 800

Elisha Curl, 20, 60, 350, 100, 260

James M. Keene, 35, -, 300, 200, 715

Isaac Smith, 65, 55, 1000, 160, 540

William C. Hair, 60, 20, 1000, 75, 1050

William Hair, 20, 20, 500, 12, 1000

William Blount, 25, 55, 500, 25, 220

Moses Smart, 50, -, 300, 25, 312

David Walker, 32, 168, 1000, 60, 526

Washington M. Ives, 30, 10, 400, 30, 420

John W. Jones, 145, 175, 2500, 500, 3010

Elias Walker, 50, 270, 2000, 200, 172

John A. W. Simons, 10, 170, 400, 100, 120

William Douberly, 50, 300, 700, 75, 630

Streaty Parker, 55, 105, 500, 40, 460

Redding Blount, 75, 85, 1000, 200, 980

Redding R. Blount, 40, 40, 600, 100, 575

Henry Pratt, 45, 35, 400, 10, 215

Arthur J. Moore, 40, 40, 400, 15, 308

Gerrett Vanzandt, 120, -, 500, 300, 980

Jacob T. Goodbread, 200, 250, 1500, 350, 1800

Samuel Barber, 50, 40, 400, 100, 500

Joseph M. Crews, 100, 60, 2000, 200, 1274

Joseph M. Hull, 150, 270, 4000, 300, 1180

John A. Johnson, 30, 80, 700, 20, 1460

Abner W. Sweat, 30, 10, 200, 100, 470

Jacob Douberly, 17, 23, 150, 40, 172

David Brown, 40, -, 500, 15, 150

John M. B. Goodbread, 15, 65, 300, 30, 246

William Kinsie, 20, 20, 200, 10, 230

Robert Brown, 150, 220, 1000, 300, 1080

Eliza Brooks, 40, 80, 1000, 100, 640

Samuel Saunders, 20, 20, 350, 30, 228

Robert Brooks, 88, 182, 1600, 240, 794

Robert S. Pryne, 33, 127, 400, 8, 596
Arthur T. Allbritton, 40, 80, 450, 30, 350
John Parrott, 22, 137, 500, 5, 340
George W. Waldron, 100, 60, 1000, 100, 1463
John M. Brannen, 30, 10, 350, 30, 740
John Peoples, 175, 200, 1500, 300, 1723
Leonard Lacke, 112, 48, 900, 75, 625
Alexander Morgan, 35, 5, 150, 50, 360
Isaac Hines, 24, 16, 162, 6, 125
Alexander Hunt, 50, 50, 80, 15, 220
Robert Sanderlin, 125, 195, 1600, 40, 865
William Cone, 150, 170, 1500, 100, 2400
John D. Pecock, 20, 20, 250, 15, 235
Jesse R. Sanderlin, 25, -, 150, 50, 485
John Fletcher, 40, 120, 800, 25, 200
Charles Fletcher, 20, -, 150, 10, 100
George Fletcher, 20, 140, 250, 40, 262
James J. Beal, 75, 50, 450, 60, 352
James L. King, 60, 100, 2000, 200, 725
Simeon Herrod, 32, -, 65, 25, 150
William Frink, 150, 10, 2000, 300, 1598
William Wilkerson, 25, 45, 275, 12, 120
William Sumerall, 115, 45, 700, 350, 1150
Isaih Clinton, 12, 28, 200, 10, 420
Edmund DeLoach, 15, -, 200, 15, 175
Carly Wiggins, 30, 130, 400, 50, 200
David Wilkerson, 80, 60, 500, 50, 915
Daniel Chavers, 20, 60, 225, 50, 240
Asa Roberts, 60, 20, 500, 120, 1320
William Roberts, 23, -, 170, 50, 358

John E. Coleman, 20, 20, 250, 15, 300
Louis Roberts, 76, -, 1500, 400, 1400
William Shirley, 20, -, 150, 20, 120
Isaih Dobson, 20, -, 400, 20, 630
Nathan M. Roberts, 20, 20, 450, 40, 825
Nathan Smith, 40, -, 250, 25, 350
Joseph Long, 14, 26, 150, 19, 275
George W. Roberts, 35, 5, 450, 100, 1090
John E. Wiggins, 10, 30, 250, 25, 350
Thomas Shirley, 20, 80, 300, 40, 400
Hansford R. Alford, 45, 35, 500, 150, 700
James R. Johnston, 20, 20, 300, 50, 230
Thomas R. Ellis, 20, 20, 200, 20, 150
William Ellis, 16, 24, 150, 40, 250
Jonathing Manning, 12, -, 100, 40, 300
William Futch Jr., 10, 30, 200, 40, 200
James Griffis, 12, 28, 200, 15, 400
Eli Griffis, 10, 30, 200, 15, 620
Richard W. Griffis, 13, 27, 250, 15, 1115
John Griffis, 25, 15, 300, 10, 1400
Samuel Varner, 15, 25, 250, 40, 200
Samuel Griffis, 7, 33, 100, 30, 150
David Browning, 15, 25, 300, 10, 340
Isaih Browning, 7, 33, 150, 10, 220
Edmund Tison, 10, 30, 150, 5, 450
Charles Griffis, 15, 25, 100, 5, 275
Morgan Prevatt, 30. 10, 400, 40, 400
Stephen J. Riggs, 10, 30, 200, 20, 360
Joseph Alvares, 15, 25, 151, 30, 850
Britton George, 30, 10, 400, 20, 600
William Johns Jr., 10, 30, 100, 5, 325
William Edwards, 30, 10, 300, 5, 200

Burrill Johns, 35, 5, 1000, 30, 700
Samuel Cruse, 12, 28, 200, 30, 600
William Johns Sr., 35, 65, 500, 30, 1300
Lucinda Johns, 20, 20, 200, 5, 380
Thomas Williams, 25, 15, 150, 10, 250
William Hernden, 30, 10, 300, 10, 160
Benjamin T. Greene, 6, -, 130, 10, 1230
Archibald H. Johns, 50, -, 1000, 12, 1945
John Robinson, 30, 10, 200, 10, 115
Isaac Ogden, 50, -, 300, 40, 1460
Drew Reddish, 15, 25, 300, 20, 1600
Francis N. Andrea, 12, 28, 500, 75, 650
Simeon D. Swift, 18, 22, 200, 100, 800
John Brown, 25, 15, 400, 40, 860
William Futch Sr., 30, 10, 340, 40, 200
Wayn Tullis, 156, 24, 160, 25, 140
John Dias, 19, 31, 150, 25, 150
Levi Pelham, 50, -, 500, 10, 700
James N. Myers, 3, 37, 200, 50, 310
Elias Wester, 25, 15, 300, 70, 520
Jacob B. Hunter, 8, 32, 300, 20, 400
William Mickle, 50, 30, 500, 10, 400
John Sweat, 75, -, 500, 200, 1250
Charles Newman, 23, 17, 250, 50, 600
Leroy Thrift, 12, 28, 200, 5, 320
William Coleman, 10, 30, 150, 30, 800
Kezie Addison, 10, 30, 200, 5, 250
John Douglass, 21, -, 200, 25, 278
William Man, 4, 36, 150, 5, 240
Benjamin Ganey, 8, 32, 200, 8, 300
David H. Jones, 25, 15, 250, 10, 375
Elisha Walker, 8, 32, 150, 35, 200
Tyri Parrish, 16, 24, 250, 100, 600
Elisabeth Powell, 10, 30, 125, 15, 350
Josiah Parrish, 12, 28, 90, 30, 250

William D. Smith, 30, 10, 600, 100, 125
Richard Tellis, 130, 330, 2000, 120, 940
Littleton Hancock, 18, 58, 500, 40, 430
James Hancock, 10, 30, 100, 10, 250
Alexander Douglass, 20, 20, 400, 40, 700
John J. Roberts, 18, 22, 250, 50, 930
John C. Winn, 25, 15, 250, 25, 640
George Varner, 40, 40, 820, 150, 780
Elisabeth Holton, 6, 34, 225, 10, 300
Charles McKiney, 55, 25, 500, 200, 1300
James F. B. McKiney, 14, 26, 200, 80, 880
John Driden, 50, 30, 350, 100, 740
Solomon Renfroe, 30, 130, 350, 100, 500
Roman Alvares, 5, 35, 125, 12, 320
Cain Strickland, 40, 160, 500, 5, 600
James O. Brooks, 5, 35, 80, 4, 350
Joel E. Renfroe, 25, 95, 500, 10, 460
Henry Sapp, 52, -, 400, 40, 740
Leaston A. Winn, 3, 37, 100, 30, 401
Moses Deas, 20, 15, 300, 25, 135
Coach R. Denison, 30, 50, 800, 50, 850
Mathew Scarborough, 12, 28, 300, 8, 160
Richard S. Mott, 45, 235, 1600, 100, 625
Robert P. Lewis, 20, 25, 400, 50, 240
Richard Ward, 11, 29, 200, 10, 80
William H. Ward, 80, -, 1500, 375, 3000
William Osteen, 8, 32, 150, 15, 500
William H. Driggers, 20, 20, 300, 10, 550
John M. Mott, 16, 24, 125, 60, 212
Solomon Godwin, 16, 24, 350, 40, 500
Jacob Godwin, 60, 100, 700, 200, 90
William Coleson, 20, 60, 500, 10, 290

Richard J. Godwin, 45, 35, 600, 125, 750

Nathan Roberts, 15, 25, 300, 50, 440

Nathaniel Roberts, 9, 31, 200, 30, 120

Henry Herrington, 55, 25, 600, 40, 600

James C. Bronson, 30, 10, 300, 40, 294

James R. Bronson, 12, 28, 100, 9, 240

Jackson Morgan, 60, 20, 1000, 200, 1140

Isaac Daniels, 30, 10, 350, 50, 600

William C. Poss (Pass), 40, -, 500, 50, 420

Joel Curry, 40, -, 700, 100, 1450

David Raulerson, 15, 25, 250, 150, 450

Jeremiah Blackwelder, 35, 5, 500, 200, 524

David Keene, 9, 31, 150, 40, 950

Wiley Keene, 6, 34, 150, 24, 560

John Harvey Jr., 30, 10, 375, 12, 380

Josiah Johnson, 12, 38, 110, 12, 400

William Hall, 10, 30, 200, 25, 400

Levi Harvey, 30, 10, 200, 12, 260

John Harvey Sr., 20, 20, 200, 10, 400

James M. Burnsed, 25, 15, 500, 200, 700

William Motes, 22, 18, 200, 50, 440

Andrew J. Harvey, 25, 15, 200, 10, 175

John J. Johnson, 40, -, 100, 10, 170

Elijah W. Greek, 20, 20, 300, 25, 80

Ivens Tumblin, 35, 5, 300, 30, 900

Archibald Hogan, 25, 15, 200, 20, 380

Thomas Thompson, 15, 25, 150, 12, 320

Errick Johnson, 12, 28, 100, 20, 360

Elisabeth Thompson, 6, 34, 90, 10, 80

Nathan Beasley, 10, 30, 100, 35, 200

Cornelius Robinson, 10, 30, 100, 4, 350

Mary Beasley, 12, 28, 40, 2, 240

William Williams, 15, 25, 100, 10, 200

Ebenezer Metts, 8, 32, 100, 50, 160

William A. Williams, 25, 15, 150, 6, 250

Roland Williams, 20, 20, 200, 5, 300

George Combs, 50, 30, 500, 75, 900

John D. Williams, 30, 10, 350, 50, 1000

David Fowler, 20, 20, 100, 50, 290

Tarlton Johns, 25, 95, 500, 50, 780

Moses Barber, 270, -, 2500, 435, 5800

Susan Barber, 50, -, 300, 20, 1100

Daniel Clifton, 10, - 150, 30, 920

Levin Tomberlin, 15, 25, 200, 30, 640

Zara Davis, 10, 30, 150, 30, 350

John Raulerson, 18, 22, 250, 10, 480

Nimrod Raulerson, 20, 20, 300, 40, 900

John Kenady, 70, 10, 300, 30, 1450

William Johns, 14, 25, 200, 100, 304

William Raulerson, 45, 115, 600, 200, 740

James Altman, 6, 34, 150, 6, 440

Jacob Raulerson, 18, 22, 150, 6, 440

West Raulerson, 10, 30, 250, 8, 720

John Cruse, 70, 10, 550, 150, 1620

Richard Harvey, 60, 20, 400, 10, 600

James Dougherty, 40, -, 800, 90, 3050

Leonard Osteen, 30, 10, 130, 5, 270

Eli Hicks Sr., 30, 10, 150, 30, 440

Eli Hicks Jr., 12, 28, 200, 25, 400

John Clifton, 40, -, 250, 40, 860

Allen Lowe, 15, 25, 200, 25, 320

Alexander Tindall, 30, 10, 200, 30, 200

Daniel J. Man, 20, 20, 200, 25, 560

William Driggers, 15, 25, 150, 15, 880

Jonas Driggers, 40, -, 400, 100, 1080

James Greene, 12, 28, 150, 40, 560

Elisha Greene, 100, 20, 500, 200, 1140

Manning Griffis, 4, 36, 150, 30, 320

John B. Whitfield, 50, -, 250, 10, 500

Louis Mattach, 260, 590, 3500, 400, 1311

Daniel T. Tresvant, 25, 40, 100, 40, 830

Thomas D. Dexter, 200, 200, 3000, 600, 1500

Nancy Sikes, 100, 200, 400, 100, 1142

James R. Robinson, 15, 65, 200, 35, 157

Nevin McLeran, 50, 90, 800, 250, 680

William G. Havens, 12, 28, 120, 10, 140

Job Manning, 50, -, 500, 25, 800

Samuel Wadsworth, 40, -, 200, 50, 530

Rhoda Dean, 100, 60, 1600, 60, 555

Joseph Durrance, 23, 17, 150, 40, 200

John W. Robinson, 40, 60, 600, 50, 350

Joseph A. Zeppers (Zefferes), 35, 5, 250, 6, 144

Solomon Zefferes, 40, 135, 700, 100, 610

Silas Overstreet, 20, 70, 250, 40, 1075

Allen Hinton, 30, -, 200, 80, 600

Edward J. Boyett, 60, 38, 800, 100, 240

Robert Greene, 30 10, 150, 100, 412

John Parker, 26, 54, 250, 25, 220

Simon Brown, 32, 8, 200, 15, 256

Green Johnston, 50, 30, 1000, 80, 880

William B. Hurst, 50, 30, 500, 25, 550

William H. Rousseau, 50, 120, 800, 200, 1214

Jesse Walder, 12, 28, 100, 10, 1550

Aldredge Wiley, 20, 20, 200, 15, 540

Mary Hardee, 50, -, 200, 10, 500

David Platt, 35, -, 300, 60, 500

Robert West, 25, 15, 200, 20, 348

Charles Lee, 25, 15, 500, 20, 400

William Platt, 20, 20, 250, 30, 420

James Polk, 20, 20, 150, 30, 250

John M. Bergan (Bergen), 40, -, 200, 50, 340

John D. Boatwright, 40, -, 150, 25, 500

Chesly J. D. Boatwright, 25, 15, 150, 30, 400

Roland Roberts, 25, 15, 100, 25, 550

Robert H. Brannon, 20, 20, 100, 5, 240

John Bevins, 10, 30, 100, 5, 200

Philip Cato, 20, 20, 200, 50, 370

Joseph Dial, 105, 527, 1000, 40, 700

John J. Taylor, 35, 200, 500, 25, 430

Mathew Mickles (Mickler), 60, 260, 1500, 100, 1100

Westbury Walker, 25, 55, 400, 35, 700

John Powell, 35, 125, 2000, 150, 800

Flemming Deas, 65, 15, 500, 40, 860

Andrew McClellan, 25, 135, 200, 10, 460

Isaih Tillis, 20, 20, 150, 10, 190

Loruns P.Brown, 12, 28, 100, 10, 160

John W. Clemments, 18, 22, 300, 50, 1068

Catharine Blackman, 17, 23, 150, 5, 192

Harris Holliman, 20, -, 100, 5, 196

Ann Bass, 20, 40, 700, 30, 136

Elijah Caraway Jr., 24, 56, 200, 20, 230

Archibald Caraway, 125, 15, 300, 35, 392

George W. French, 12, 28, 100, 5, 132

Elijah Caraway Sr., 100. 60, 700, 100, 746

Marmaduke Blackburn, 55, 25, 150, 40, 640

Jasper Evers (Evens), 35, 5, 200, 5, 710

Green Williams, 15, 25, 125, 15, 130

Thomas Evers, 20, 20, 100, 15, 290

Benjamin King, 25, 95, 250, 15, 140

Judge R. Hodges, 80, 240, 1000, 200, 400

Blackston Ellis, 130, 230, 1200, 150, 720

Cintha Sweney, 10, 30, 150, 5, 108

Levi Johns, 25, 15, 300, 50, 700

Jefferson Williams, 30, 10, 200, 10, 476

Richard Woodward, 12, 28, 200, 20, 350

William Holmes, 20, 20, 200, 15, 292

Sampson Carver, 40, 40, 400, 100, 870

James Bryant, 40, -, 150, 40, 700

John W. Gilespie, 60, 20, 600, 75, 450

William Carver, 50, 70, 500, 50, 500

William T. Niblack, 70, 10, 350, 60, 1230

Ann Niblack, 30, 10, 175, 30, 1170

Edgar Collins, 45, 35, 800, 140, 700

William Cason, 250, 1752, 6000, 100, 400

James Goff, 10, -, 100, 10, 180

Charles H. B. Collins, 65, 15, 500, 75, 500

Acher Goff, 20, 20, 150, 20, 540

Redding Long, 20, 20, 150, 40, 700

David Cannon, 17, 23, 150, 50, 500

Nathan Smith, 8, 32, 400, 4, 270

James Osteen, 15, 25, 300, 50, 500

George E. McClellan, 120, 370, 3000, 200, 1200

James Deas, 20, 20, 160, 20, 270

Thomas Gaskins, 40, 160, 1200, 40, 700

Ransom Parrish, 20, 20, 200, 40, 525

Oden Parish, 10, 30, 100, 10, 150

Thomal S. Hawkins, 20, 100, 200, 100, 300

Thomas Hawkins, 125, 275, 1000, 50, 1300

John W. Hawkins, 5, 35, 100, 50, 360

Richard Clark, 6, 74, 160, 10, 226

Jacob Davis, 8, 32, 140, 35, 200

John Tillet, 9, 31, 70, 30, 140

George Keene, 113, 27, 200, 15, 575

Shaderick Hancock, 85, 225, 1000, 100, 1700

Herman Lane, 13, 27, 115, 5, 200

William Townsend, 120, 160, 4500, 200, 1810

Elias Osteen, 40, -, 200, 40, 600

Giles U. Ellis, 20, 60, 300, 30, 430

Theophilus Weeks, 100, 566, 4000, 400, 1700

Barnett C Weeks, 20, 60, 300, 30, 430

Henry C. Willson, 35, 45, 250, 40, 433

John Vanzandt, 12, 38, 300, 35, 324

Bartholome Cason, 15, 25, 300, 6, 250

Henry Vanzandt, 12, 68, 150, 5, 361

Reuben Osteen Sr., 70, 10, 200, 50, 844

Reuben Osteen Jr., 15, 25, 200, 30, 748

Jacob Williams, 25, 15, 100, 20, 540

Isaac Moody, 12, 28, 75, 5, 100

Roland Thomas, 20, 60, 500, 25, 450

James Cason, 10, 30, 300, 10, 276

John Mathews, 50, 190, 800, 200, 957

Benjamin Moody, 200, 240, 1000, 200, 1674

Louis Tucker, 11, 29, 75, 40, 203

Daniel Hurst, 40, 40, 800, 50, 471

William Scott, 60, 60, 1400, 150, 2407

Benjamin Lane, 13, -, 100, 10, 116

Samuel H. Worthington, 15, 145, 500, 50, 330

Samuel Worthington, 12, 28, 400, 10, 260

John L. Moody, 12, 28, 100, 5, 394
William C. Parrish, 25, 15, 200, 5, 290
Owen Revels, 30, 10, 300, 50, 250
John S. Parrish, 10, 30, 150, 10, 450
Jourdon L. Gaskins, 20, 20, 300, 40, 1050
Joseph M. Brooks, 10, 30, 100, 10, 418
Allen Hazell, 8, 32, 100, 10, 200
John Prevatt, 20, 20, 150, 25, 786
Reuben Prevatt, 8, 32, 60, 5, 118
John M. Prevatt, 35, 165, 600, 40, 580
William Barco, 26, 17, 600, 75, 358
Daniel Barco, 25, 15, 450, 30, 313
Thomas Kelly, 30, 10, 300, 30, 180
Elijah H. Mattox, 200, 80, 600, 40, 556
Eliza M. Stewart, 226, 300, 1500, 400, 1650
Daniel D. Greene, 11, 29, 150, 30, 216
Stephen Barco, 18, 22, 200, 100, 604
Joseph Barco, 15, -, 100, 10, 194
Edward Williams, 20, 12, 150, 30, 380
John Combs, 10, -, 125, 5, 68
Ezekiel Weeks, 70, 206, 1500, 70, 678
Silas Weeks, 20, 20, 100, 10, 360
Sarah Powell, 10, 30, 100, 5, 224
James Powell, 50, 110, 500, 40, 300
William Beasley, 30, 460, 1150, 20, 577
Temperance Mickles (Mickler), 40, -, 100, 10, 100
Rebecca Charles, 85, 418, 1000, 150, 3780

John W. Herring, 60, 100, 800, 50, 280
Daniel Hall, 15, 25, 175, 50, 900
James D. Lee, 10, 30, 175, 10, 100
Dempsey Sawyer, 8, 32, 160, 10, 170
Isaac Walker, 15, 25, 100, 40, 350
Sarah Edwards, 15, 25, 200, 15, 300
John Edwards, 10, 30, 120, 10, 250
William Edwards, 20, 20, 300, 15, 360
William Roland, 15, 25, 600, 25, 950
Albert Peterson, 10, 30, 125, 5, 400
Robert Rollins, 30, 50, 400, 30, 160
William Riggs, 25, 15, 200, 10, 100
Alexander Holmes, 80, -, 500, 40, 445
Thomas Watters, 8, 32, 125, 10, 108
Harrison Raulerson, 70, 10, 400, 25, 2330
John Burnell, 30, 10, 400, 25, 1150
Mahan Driggers, 14, 26, 100, 20, 374
David Waldron, 25, 55, 200, 100, 700
Thomas H. Albritton, 80, -, 200, 25, 600
David Long, 10, 70, 175, 20, 160
Jessee Long, 25, 35, 150, 25, 400
John Rutledge, 12, 28, 150, 20, 300
William Brown, 15, 25, 150, 10, 325
John Cason, 8, 32, 80, 30, 450
Simeon Sheffield, 30, 10, 200, 30, 400
William H. Parrish, 35, 5, 400, 50, 750
Allen Thomas, 23, 47, 350, 40, 533

Dade County, Florida
1850 Agricultural Census

The University of North Carolina at Chapel Hill filmed the 1850 agricultural census for Dade County from originals at the Florida State University under a grant from the National Science Foundation in 1963.

Columns 1, 2, 3, 4, 5, and 13 represent the following information on the census:
1. Name of Owner, Agent or Manager of Farm
2. Acres of Improved Land
3. Acres of Unimproved Land
4. Cash Value of the Farm
5. Value of Farming Implements and Machinery
13. Value of Livestock

James Wright, 10, -, 500, 100, 175
Ed Baizely (Baizeby), 4, 36, 800, 100, 25
Jos. H. Sanders, 4, -, 100, -, 10
Richd. Russell, 15, -, 300, -, -
John Curry, 8, -, 200, -, -
Asa Wetherford, 10, -, 200, -, -

Duval County, Florida
1850 Agricultural Census

The University of North Carolina at Chapel Hill filmed the 1850 agricultural census for Duval County from originals at the Florida State University under a grant from the National Science Foundation in 1963.

Columns 1, 2, 3, 4, 5, and 13 represent the following information on the census:
1. Name of Owner, Agent or Manager of Farm
2. Acres of Improved Land
3. Acres of Unimproved Land
4. Cash Value of the Farm
5. Value of Farming Implements and Machinery
13. Value of Livestock

Matthew Butler, 25, -, 250, 20, 265
Jesse Bennett, 20, -, 125, 10, 110
Hiram Bennett, 14, -, 150, 10, 462
Jackson Sullivant, 8, -, 100, 10, 190
James Knight, 30, -, 150, 10, 395
John Walker, 13, -, 100, 5, 108
William Knight, 16, -, 200, 8, 460
Darling C. Prescott, 35, 50, 2000, 150, 425
Moses Prescott, 40, -, 200, 10, 662
William Long, 4, -, 150, 5, 380
Elias Padgett, 10, -, 200, 5, 223
William Padgett, 12, -, 200, 10, 505
Nelson Ferris, 8, -, 150, 5, 90
Robert B. Sullivant, 19, -, 400, 15, 200
Silvey Wilford, 14, -, 100, 5, 356
Hugh Tyler, 4,-, 150, 3, 40
Cullen Rowles, 12, -, 150, 20, 175
Winna Moore, 25, -, 150, 5, 130
John Moore, 12, -, 160, 5, 1240
John G. Caldwell, 15, -, 100, 2, 205
William Silcox, 15, -, 150, 10, 150
William Dessance, 10, -, 200, 15, 445
Michael Sloan, 10, 20, 200, 35, 100
Charles J. Schornhurst, 5, 28, 300, 10, 140
George Hartley, 25, 359, 1000, 150, 686
Josiah Hagens, 14, 36, 125, 10, 150
Daniel Sanders, 5, 35, 125, 10, 110

Thomas Gardner, 5, -, 100, 5, 50
Joseph Marten, 10, -, 100, 5, 164
George A. Hartley, 6, 290, 300, 10, 190
Sarah Hagen, 6, 14, 200, 5, 200
Nathaniel Hartley, 20, 20, 1000, 150, 1000
William J. Hartley, 6, 34, 150, 10, 175
Michael Hartley, 22, 35, 200, 40, 500
John L. Hamilton, 14, 36, 400, 100, 610
Sarah Acosta, 8, 165, 300, 5, 20
Sarah Carter, 4, 100, 100, -, 280
Jane H. Hanley, 10, -, 150, 15, 120
Andrew Jones, 8, -, 175, 10, 150
Lewis Brown, 6, 34, 200, 15, 200
Henry Kura, 3, 3, 200, 10, 100
John Schaffler, 30, 300, 1000, 20, 900
Samuel T. Thompson, -, -, -, -, 180
James Fagen, 12, 90, 1000, 15, 300
Isaac Narn, 6, 230, 500, 15, 120
James Kellum, -, -, -, -, 75
Moses Joiner, 6, -, 100, 5, 180
George Branning, 200, 1700, 4000, 1000, 6315
Isaac Green, 14, -, 200, 5, 580
John Sumerlin, 120, 140, 2000, 200, 1310
Nancy Lee, 7, -, 150, 15, 230

John Roberts, 5, -, 150, 10, 37
William Roberts, 8, -, 200, 12, 140
John Black, rented, rented, rented, 15, 300
Lervin Johnson, rented, rented, rented, -, 90
Thomas Hendrick, 120, 640, 800, 75, 2130
Jonathan Knight, 80, 137, 1500, 35, 700
Elijah Blitch, 60, 500, 700, 100, 658
Osias Buddington, 45, 500, 5000, 200, 720
John G. Smith, 25, 160, 2600, 70, 213
Moses Prescott, rented, rented, rented, 10, 125
John H. Wheeler, 18, 120, 1500, 100, 190
Jane A. Snowden, rented, rented, rented, -, 140
James J. Pool, 18, -, 150, 15, 65
Richard Long, rented, rented, rented, -, 75
Benjamin Frisbee, 20, 40, 600, 75, 32
Nathaniel Pimento, 9, -, 150, 10, 100
William W. Johnson, 4, -, 150, 5, 75
Bird Knowles, rented, rented, rented, 10, 55
Jeremiah Halsey, 15, 25, 100, 10, 360
Burrett Yates, 12, -, 150, 5, 137
Anthony Folona, rented, rented, rented, 5, 130
Stephen Padgett, rented, rented, rented, -, 74
William Stott, 3, -, 200, -, 414
Thomas Dillberry, 50, 130, 1500, 75, 1350
William Taraters, 25, 31, 300, 60, 400
Fredric Barthlow, 8, 32, 250, 5 , 2
Andrew J. Philips, 60, 1131 500, 100, 789
Isaac Varn Jr., 40, 120, 350, 30, 300

Thomas Smith, rented, rented, rented, 15, 250
Henry Silcox, rented, rented, rented, 10, 290
William S. Bardin, 20, 80, 300, 40, 250
James Bardin, 10, -, 150, 25, 150
J. Harderbrook, 18, -, 200, 10, 256
Jesse Lee, 1, -, 50, -, -
Abel Jones, 8, -, 150, 5, 8
C. Braddock, 30, 10, 400, 20, 770
William Harris, rented, rented, rented, -, -
Francis Wood, rented, rented, rented, -, 700
James Richardson, 15, -, 150, 10, 460
Ephraim Conway, 5, 9, 200, 5, 430
Mesheck Burney, rented, rented, rented, 10, 290
John H. McIntosh, 800, 6000, 30000, 2000, 5900
Henry Yates, rented, rented, rented, 8, 90
George Tippins, 40, 80, 700, 50, 302
William H. G. Saunders, 100, 20, 1000, 25, 900
Alfred Taylor, -, -, -, 25, 1111
George Flemming, 80, 270, 2000, 150, 580
Charles Byrne, 100, 200, 4000, 100, 1000
Isiah D. Hart, 120, 1000, 5000, 200, 1000
William Sedgwick, 10, -, 100, -, 100
Arthur M. Read, 15, -, -, -, 375
Thomas Douglass, 2, -, -, -, 250
Mark Butts, -, -, -, -, 200
Ebenezar Ereleth, 200, 100, 1500, -, 160
Moses Curry, 20, 980, 225, 100, -
Henry D. Holland, -, -, -, -, 125
Jacob Joreman, 250, 800, 800, -, 15
Thomas Ledwith, 5, 10, 150, 10, 900
James M. Daniels, -, -, -, -, 1680
William P. Desaces, -, -, -, -, 80

Samuel L. Burritt, -, -, -, -, 320
Maria Doggett, 7, 10, 1000, -, 70
George Trosts, -, -, -, -, 10
Stephen D. Fernandes, -, 5850, - -, 110
Anthony Canova, 20, 20, 150, -, -
Thomas W. Jones, 12, 858, 520, 100, 460
James W. Bryant, -, -, -, -, 210
Mary Hall, -, -, -, -, 250
Jesse Townsend, -, -, -, -, 150
Jace W. Hand, -, -, -, -, 130
Lorenza Allen, 6, 33, 200, 20, 150
Jacon Moody, 2, -, -, -, 200
Uriah Roberts, 5, -, 100, 10, 140
Alexander Roberts, 12, -, 150, 30, 415
Elias G. Jaudon, 400, 500, 1000, 400, 1610
Willoughby Tucker, 6, -, -, 10, 85
Thomas J. Jones, 30, 10, 250, 30, 242
Joseph Strickland, -, -, -, -, -
James Winter, 100, 700, 5000, 200, 418
Hilliard Jones, 7, -, 600, 10, 632
John A. Hogans, 7, 150, 200, 50, 760
Francis S. Hurdnall, 20, 40, 800, 60, 780
Mary Hagan, 7, 43, 200, 20, 234
Isaac Bush, 20, 40, 100, 50, 450
James A. Pickett, 10, 380, 900, 10, 720
John Harroll, -, -, -, -, 130
Lewis Ryals, -, -, -, -, 75
James Cammel, -, -, -, -, 50
John Sadbery, 5, 35, 100, 5, 60
Patric Price, 8, 32, 250, 5, 410
William Hammond, -, -, -, 5, 100
Elijah Higgenbothan, 40, 200, 900, 40, 1220
David Tanner, 40, -, 150, 20, 505
Isaac Wingate, 50, -, 120, 20, 50
James Higgenbothan, 60, -, 300, 20, 320

Elijah Higgenbothan, 30, -, 100, 10, 350
William Stone, 18, -, 150, 12, 400
David Stone, 2, -, 50, 5, 130
James Wingate, 3, -, 75, 6, 190
John Harroll, 6, -, 60, 5, 120
Nelly Buckler, 3, -, 60, 5, 112
Martha Huffin, 6, -, 50, 2, 172
John Price Jr., 100, 111, 1800, 275, 1360
Henry Dias, -, -, -, -, 130
John A. Jones, 300, 2200, 4725, 200, 2500
William C. Tison, 10, -, 200, 20, 126
John Roberts, 22, 98, 400, 125, 826
Isaac Roberts -, -, -, -, 175
Jefferson Higgenbothan, 30, -, 300, 10, 325
Seaborn Edge, 10, -, 150, 5, 213
Tony Mayo, 3, -, -, -, 360
Jacob Whiteacker, -, -, -, -, -
William Haddock, 80, 220, 610, 75, 1300
Arhur Burney, -, -, -, -, -,
James McDonald, -, -, -, -, 516
Margaret Suares, 2, -, 100, 5, 120
Emanuel Mott, 8, -, 150, 6, 110
Thomas Warren, 6, -, 100, 5, 150
John D. Rantin, 7, -, 150, 6, 80
Thomas Suares, 30, -, 300, 30, 1060
John Huffin, 20, -, 200, 25, 288
Seymore Pickett, 20, -, 130, 10, 1023
Daniel Hogan, 6, -, 100, 20, 344
Joseph Eneevs(Enecos), 20, -, 200, 40, 547
George Stone, 20, 120, 600, 50, 110
Daniel S. Gardner, 15, 65, 500, 50, 85
Susan Bryant, 40, -, -, 10, 170
Miles Price, 200, 1300, 3000, 200, 1862
Henry O'Neil, 20, -, 200, 30, 1025
Francis Riggs, 10, -, 400, 20, 522
Daniel H. Youngblood, 12, -, 150, -, 600

Robert Bigelow, 200, 200, 5000, 50, 240

Albert G. Philips, 200, 1050, 6000, 600, 1310

Henry Clifton, 12, -, 200, 15, 150

James T. McCormick, 18, -, -, 10, 70

Sarah A. Driggers, 6, -, -, -, 40

James McCormick, 26, -, 300, 15, 732

Farguahar Bethune, 60, 250, 1500, 100, 100

John A. Taylor, 25, 475, 1000, 20, 890

James Silcox, 7, 33, 250, 15, 960

Sarah P. Ferris, 200, 3000, 10000, 100, 770

Edward G. Gilchrist, rented, rented, rented, -, 110

John Brady, 4, 36, 100, -, 100

Uriah Bowden, 5, 65, 200, 25, 400

John Brantley, 30, 37, 2000, 10, 260

Josiah Strickland, 7, -, 150, 10, 70

John S. Sammis, 300, -, 20000, 150, 712

Martha Baxter, 150, -, 10000, -, 150

Charles McNeil, -, -, -, -, 450

Henry R. Saddler, 700, 4500, 10000, 500, 3200

John Broward Jr., 30, 100, 1000, 50, 2850

Robert Turner, 15, -, 200, 20, 512

Robert Robertson, 15, -, 200, 30, 270

David Stratton, 8, -, 100, 10, 140

Thomas Stratton, 5, -, 50, -, 45

John J. Tucker, 6, -, 75, 10, 400

David Turner, 80, 400, 500, 50, 640

Nancy Nelson, rented, rented, rented, -, 210

John Broward, 150, 150000, 20000, 500, 2708

Luther H. Tison, 50, -, 100, 60, 400

John N. Minor, -, -, -, -, 120

William Eubanks, 130, 670, 1600, 200, 640

James Eubanks, 30, -, 100, 50, 570

William Braddock, 40, 600, 1000, 125, 400

David Oglesbee, 30, 270, 300, 20, 825

Thomas Sterrett, -, -, -, -, 250

Jefferson P. Belknap, 30, -, 200, 50, 310

James Gillame, 15, -, 100, 5, 40

Francis A. Johnson, 24, -, 100, 50, 540

William Turner, 15, -, 50, 10, 25

Felix Giger, 20, -, 200, 200, 1380

Allen Giger, 25, -, 300, 150, 1014

George Pendarvis, 20, -, 300, 300, 1330

Charles Broward Sr., 150, 650, 2000, 200, 1266

James Flinn, 20, -, -, 15, 200

Michael H. Hartley, 25, 340, 400, 15, 130

James Plummer, 200, 21, 200, 300, 1070

Alexander Plummer, 12, 188, 800, 50, 158

William Plummer, 12, 35, 500, 40, 230

Elisabeth Alger, 15, -, 300, 10, 135

Charles H. Dibble, 10, 30, 200, 50, 270

Chandler S. Cenery, 40, 340, 1500, 75, 233

John M. T. Bowden, 30, 350, 1000, 100, 500

Thomas Hagen, 15, -, 200, 60, 332

Calvin Read, 15, 200, 3000, 100, 231

Charles Read, 8, 4, 300, -, 16

Henry F. Hartly, 4, -, 60, -, 49

Mary Ann James, 15, 110, 150, 15, 93

John W. Brady, 40, -, 300, 20, 350

John A. Summerall, 15, 15, 300, 40, 615

Joseph C. Summerall, 15, 15, 150, 40, 50

William Loften, 12, 288, 600, 20, 80

Elijah Petty, 11, 29, 400, 10, 306

William M. Hagen, 10, 15, 200, 5, 190

James Weeks, 6, -, 150, 5, 300

William Wilson, 12, -, 200, 40, 320

William Thomas, 35, 10, 700, 75, 650

Benjamin Higgenbothan, 20, 60, 300, 40, 470

Cator Munroe, rented, rented, rented, -, 40

James White, 3, -, 40, 5, -

Jesse Wilson, 20, -, 150, 25, 510

James Booth, 15, 25, 100, 10, 180

James H. Madison, rented, rented, rented, 5, 249

Abraham Bessant, 100, 450, 2000, 220, 1700

William Branning (Browning), 20, 20, 400, 40, 138

James Odum, rented, rented, rented, -, 1035

James Baldwin, 22, -, 500, 50, 950

John Philips, 10, -, 400, 20, 630

Asa Emanuel, 12, -, 200, 10, 50

James McCleland, 20, -, 100, 10, 175

James Yates, 40, -, 500, 75, 460

George Chauncey, 3, -, 150, 5, 80

Needham Yates, 20, -, 300, 20, 1246

Lewis Oglesbee Jr., 25, -, 200, 15, 105

Lewis Oglesbee Sr., 6, -, 75, 5, 125

William Rothe, 6, -, 100, 10, 40

William Flinn, 1, -, 50, 10, -

James Burns, 14, -, 250, 10, 418

William H. Thomas, 15, 65, 500, 40, 810

Milton Haynes, 300, 2700, 5000, 500, 600

David P. Houston, 80, 95, 800, 20, 460

William G. Christopher, 80, 95, 1200, 30, 776

Edward Hopkins, 700, 3000, 7000, 400, 500

David L. Palmer, 300, 2700, 6000, 500, 460

Kingsley B. Gibbs, 650, 2500, 15000, 5000, 1000

John Houston, 300, 1400, 12000, 300, 1120

John Richardson, 15, 150, 1000, 15, 500

Lewis Christopher, 400, 750, 9000, 250, 300

Escambia County, Florida
1850 Agricultural Census

The University of North Carolina at Chapel Hill filmed the 1850 agricultural census for Escambia County from originals at the Florida State University under a grant from the National Science Foundation in 1963.

Columns 1, 2, 3, 4, 5, and 13 represent the following information on the census:
1. Name of Owner, Agent or Manager of Farm
2. Acres of Improved Land
3. Acres of Unimproved Land
4. Cash Value of the Farm
5. Value of Farming Implements and Machinery
13. Value of Livestock

Dillon Jordan, 200, -, 400, -, 300
Ambrose Jones, -, -, -, -, 100
Jesse Pritchett, 160, -, 500, -, 600
Wm. H. Baker, -, -, -, -, 500
Gonzalez & Bonifay (Gonzalez E. Bonifay),
James Neal, 10, 145, 1000, 150, 200
Daniel Shepard, 20, 87, 1000, 100, 900
William Waters, 10, -, 800, 100, 900
Isrial Blanchard, 5, -, 500, 25, 200
Clemson Hall, 16, -, 100, 20, 750
Micajah Andrews, 15, -, 250, 100, 35
John Milsted, -, -, -, -, 300
Delijah Lawson, 25, 55, 700, 30, 250
John Morgan, 75, 15, 2000, 50, 2500
Wm. Tenant, 40, -, 300, 50, 1000
John B. Tenant, 22, -, 100, 50, 500
Reuben Bowman, 15, 65, 200, 75, 2000
Wm. H. Fletcher, 20, -, 150, 70, 350
Andrew J. Hall, 8, -, 100, 50, 350
Lewis B. Appelegate, 30, 520, 1500, 50, 250
John Johnson, 30, 712, 1000, 100, 700
James Miles, 20, 40, 300, 50, 800
John N. Williams, 12, -, 200, 25, 130

Burgess Miles, 10, -, 100, 10, 300
Henry R. Gaylor, 8, 32, 200, 75, 900
Shephard Gaylor, 20, 20, 350, 100, 800
John Lee, 30, -, 310, 100, 200
Washington Emerson, 10, -, 100, 10, 100
Robt. W. Bowman, 30, -, 200, 50, 1200
Wm. H. Leatherwood, 60, -, 200, 50, 600
Benj. Nettles, 25, -, 200, 75, 200
Abm. Milato, 100, 4590, 6000, 150, 10000
Aaron Milsted, 50, -, 600, 75, 2500
Thos. Cooper, 10, 630, 2000, 50, 520
Valentine Fillingan, 5, -, 150, 30, 300
James McCluskey, 10, -, 100, 30, 500
Joseph Bates, 7, -, 150, 30, 500
Bernard Collins, 30, 90, 500, 25, 200
Wm. H. Chase, 20, 130, 3000, 20, 800
David Williams, 10, -, 3000, -, 4000
Elijah Herndon, 4, -, 100, 10, 200

Franklin County, Florida
1850 Agricultural Census

The University of North Carolina at Chapel Hill filmed the 1850 agricultural census for Franklin County from originals at the Florida State University under a grant from the National Science Foundation in 1963.

Columns 1, 2, 3, 4, 5, and 13 represent the following information on the census:
1. Name of Owner, Agent or Manager of Farm
2. Acres of Improved Land
3. Acres of Unimproved Land
4. Cash Value of the Farm
5. Value of Farming Implements and Machinery
13. Value of Livestock

Patrick Grady, -, 186000, -, -, 100
James Flynn, -, -, -, -, 5
John Mahoney, -, -, -, -, 5
Philip Laurence, -, -, -, -, 5
Carson Allers, -, -, -, -, 10
Owen McConnel, -, -, -, -, 5
Cornelius Grady, -, -, -, -, 50
T. D. Gibson, -, -, -, -, 100
William Petry, -, -, -, -, 10
James P. Posm, -, -, -, -, 8
William Blount, -, -, -, -, 350
F. M. Bryan, -, -, -, -, 200
Nancy Nickles, -, -, -, -, 200
Martin Corigan, -, -, -, -, 100
John Coupe, -, -, -, -, 10
Robert Myers, -, -, -, -, 30
C. J. Shepard, -, -, -, -, 130
Catharine Donahoo, -, -, -, -, 15
John Coupe, -, -, -, -, 20
George J. Broughton, -, -, -, -, 155
Phineas Laprade, -, -, -, -, 10
William McIntosh, -, -, -, -, 175
P. W. Collins, -, -, -, -, 100
John G. Ruan, -, -, -, -, 300
Dared G. Raney, -, -, -, -, 71
S. S. Hoge, -, -, -, -, 600
J. W. Rinaldi, -, -, -, -, 110
William A. McKenzie, -, -, -, -, 110
John Crooken(Crooker), -, -, -, -, 100
John Garvin, -, -, -, -, 300
Richard Coleman, -, -, -, -, 50

John Dryer, -, -, -, -, 600
Millage Stewart, -, -, -, -, 90
Richard Parker, -, -, -, -, 20
Edward Austin, -, -, -, -, 15
Lear Roderick, -, -, -, -, 100
J. W. Fountain, -, -, -, -, 40
B. S. Hawley, -, -, -, -, 150
Anthony Murrey, -, -, -, -, 15
A. G. Semmes, -, -, -, -, 230
T. L. Mitchell, -, 286000, -, -, 900
Clinton Thigpin, -, -, -, -, 500
George Sinclair, -, -, -, -, 450
W. H. Long, -, -, -, -, 100
Wm. Valleau, -, -, -, -, 30
S. W. Spencer, -, -, -, -, 115
L. H. Goodman, -, -, -, -, 15
N. Baker, -, -, -, -, 15
A. T. Bennett, -, -, -, -, 225
A. N. McKay, -, -, -, -, 400
P. C. Kain, -, -, -, -, 50
C. H. Marthes, -, -, -, -, 60
T. Orman, -, -, -, -, 100
Wm. T. Wood, -, -, -, -, 300
Benj. Lucas, -, -, -, -, 100
R. H. Goodtell, -, -, -, -, 40
County Franklin, -, -, -, -, 100
Wm. C. McNeal, -, -, -, -, 1300
Alexander Yant, -, -, -, -, 1000
Henry Cough, -, -, -, -, 330
J. Wilson, -, -, -, -, 250
John Ramsey, -, -, -, -, 100

Wm. Marr, -, -, -, -, 20
George McLeghorn, -, -, -, -, 600
John Younger, -, -, -, -, 120
Miles Fields, -, -, -, -, 220
Jeremiah Tate, -, -, -, -, 400
Jesse F. Potts, -, -, -, -, 40
H. F. Simmons, -, -, -, -, 266
H. D. Darden, -, -, -, -, 700

C. G. Holmes, -, -, -, -, 200
H. R. Taylor, -, -, -, -, 60
J. Day & Co., -, -, -, -, 650
Wm. H. Younger, -, -, -, -, 100
N. P. Deblois, -, -, -, -, 300
G. S. Hawkins, -, -, -, -, 300
J. J. Gripper (Griffin), -, -, -, -, 100

Gadsden County, Florida
1850 Agricultural Census

The University of North Carolina at Chapel Hill filmed the 1850 agricultural census for Gadsden County from originals at the Florida State University under a grant from the National Science Foundation in 1963.

Columns 1, 2, 3, 4, 5, and 13 represent the following information on the census:
1. Name of Owner, Agent or Manager of Farm
2. Acres of Improved Land
3. Acres of Unimproved Land
4. Cash Value of the Farm
5. Value of Farming Implements and Machinery
13. Value of Livestock

John G. Gunn, 80, 240, 2000, 50, 500

Thomas H. Condry, 110, 120, 2000, 600, 920

Hiram Munroe, 900, 1255, 12000, 700, 1900

Arthur J. Foreman, 900, 1400, 18000, 1000, 3000

Joseph Austin, 500, 500, 7000, 200, 1700

James Whaley, 150, 250, 1000, 300, 650

Thomas J. Potts, 140, 300, 1000, 100, 360

William T. Stockton, 50, 5, 500, 220, 770

Jacob Bradwell, 200, 600, 6000, 500, 1225

William Smith, 10, -, 200, 300, 170

Benjamin Smith, 90, 110, 1400, 150, 700

William S. Gunn, 400, 880, 6400, 600, 1100

Bolling Baker, 200, 200, 2500, 50, 500

Martin Dolan, 40, 40, 160, 50, 150

Harris T. Wyatt, 200, 390, 4500, 500, 750

Lucy Lambert, 100, 60, 500, 110, 300

Eligah S. Sheppard, 200, 100, 1000, 400, 1000

Roderick K. Shaw, 200, 1000, 5000, 500, 1050

James M. Gilchrist, 140, 200, 2000, 400, 500

Carter T. Baughan, 450, 220, 3000, 300, 750

William J. Attwater, 1120, 40, 2000, 300, 1050

Charles N. Dupont, 200, 1800, 15000, 800, 3000

Alexander McIver, 200, 200, 3000, 300, 850

John McDonald, 30, 130, 500, 50, 425

Murdock Shaw, 100, 180, 2500, 200, 450

John Shaw, 50, 210, 1300, 75, 200

Isaac R. Harris, 200, 800, 5000, 600, 1600

David L. White, 250, 650, 6000, 1000, 1000

Willoughby S. Gregory, 350, 450, 7000, 600, 1350

William Forbes, 250, 750, 9000, 500, 1300

John W. Poindexter, 350, 450, 7000, 500, 1500

Henry W. Brown, 400, 350, 5000, 600, 1450

William Booth, 800, 400, 6000, 650, 400

John Pittman, 15, 285, 1000, 50, 400

Isaac Wilson, 80, 80, 500, 175, 400

Thomas Munroe, 1200, 1010, 15000, 1100, 2980

Elisha Strickland, 8, 92, 200, 40, 100

Adam Hardin, 25, 175, 500, 60, 330

Thomas D. Wilson, 20, 20, 200, 30, 250

William H. Brown, 50, 310, 500, 110, 300

William Hagood, 60, 50, 600, 250, 425

Archibald Smith, 180, 620, 4000, 600, 1860

William Jeter, 125, 1950, 2250, 310, 900

Jean H. Verdier, 80, 400, 6000, 1000, 1900

Clinton Beckwith, 40, 160, 350, 50, 220

William Nunan, 100, 220, 2250, 300, 650

Charles S. Sibley, 600, 800, 25000, 1000, 2500

John McLaughlin, 175, 305, 3500 600, 1200

Cullin Edwards, 30, 90, 600, 50, 550

Joshua Sheppard, 100, 106, 1000, 150, 650

William Kelly, 40, 120, 600, 50, 250

Robert Clark, 70, 130, 500, 150, 400

James Sherrod, 20, 155, 300, 50, 400

William Mercer, 50, 100, 200, 40, 400

Daniel Shaw, 30, 130, 300, 75, 570

Elbert Long, 70, 130, 1000, 50, 500

Philip A. Stockton, 80, 40, 1200, 175, 500

Margaret Smith, 200, 60, 1000, 100, 900

David S. McBride, 200, 400, 3300, 300, 2000

James Wylie, 125, 70, 3000, 250, 600

Jesse Williams, 30, 290, 1000, 50, 250

Edwin Holt, 6, 36, 50, 20, 150

Edward Thomas, 100, 500, 3000, 250, 900

David Sheppard, 50, 190, 1000, 50, 600

George Love, 100, 300, 3000, 200, 500

William Maks, 30, 30, 500, 20, 200

James Fellizune, 1700, 700, 18000, 1000, 240

Joshua Drake, 60, 50, 400, 50, 800

Whitehurst Hawkins, 60, 260, 1600, 100, 1000

John Colson, 300, 320, 3000, 200, 1000

John Wymer, 420, 455, 2500, 200, 700

Henry Lipford, 400, 523, 8000, 750, 1500

John McLilly, 100, 300, 1000, 75, 300

Robert L. Harrison, 400, 520, 5000, 300, 1500

James Strange, 40, 360, 1500, 75, 450

Hilliard Derrick, 70, 430, 2500, 125, 100

Madison Herrington, 15, 135, 750, 40, 300

Ira Carner(Carra), 20, 30, 275, 50, 350

William Mann, 12, 75, 250, 25, 250

Daniel Lustin (Lustie), 70, 90, 1500, 30, 450

Thomas Parrish, 30, 50, 300, 25, 120

Marty Martin, 100, 220, 2000, 150, 650

Thomas Cain, 31, 100, 1000, 25, 100

James J. Wood, 45, 135, 1500, 75, 250

Gasper V. Sivert, 125, 315, 3000, 150, 1000

Hadley Hinson, 115, 300, 2000, 150, 800

John Smith Sr., 250, 300, 8000, 600, 1650

Hiram Edwards, 40, 40, 900, 50, 400

David Holloman, 125, 235, 5000, 50, 1200

Daniel Sove (Love), 140, 320, 5000, 125, 1750

Joseph Chauncey, 40, 160, 800, 25, 600

Anne Burnley, 200, 60, 5000, 150, 900

Archibald McPhoul, 70, 120, 800, 30, 600

Elisha Bradshaw, 140, 140, 1000, 75, 600

Daniel Hinson, 300, 780, 4000, 300, 1000

John T. Smith, 25, 132, 600, 25, 250

Jane Parramore, 160, 120, 1100, 125, 1000

John D. Cowart, 200, 400, 3000, 140, 950

Thomas Holliday, 160, 240, 6000, 500, 900

John Robinson, 17, 143, 300, 25, 125

John McNair, 40, 60, 500, 30, 1600

John D. Rawls, 250, 400, 5000, 500, 1400

Latinus Armistead, 25, 1400, 7000, 25, 150

Sarah J. Lines, 800, 900, 15000, 1500, 300

Henry Gee, 400, 800, 15000, 1000, 3500

Nathaniel Zeigler, 500, 500, 12000, 1000, 3000

Benjamin Watson, 24, 76, 1000, 50, 400

Samuel Colluck, 275, 205, 4000, 140, 1250

Andrew McCoy, 20, 280, 1500, 25, 800

James Mayo, 50, 350, 5000, 125, 750

Edmund Gregory, 150, 240, 2500, 100, 900

James Meachem, 160, -, 800, 75, 400

Banks Meachem, 390, 390, 5000, 350, 1500

Robert F. Jones, 25, 55, 800, 25, 475

John McDougall, 50, 200, 500, 25, 600

Neill Black, 120, 80,750, 40, 500

Miles Johnson, 150, 430, 5500, 125, 1400

John Johnson, 150, 430, 5500, 110, 1500

John Howard, 70, 76, 800, 75, 400

Daniel Buie, 90, 310, 2000, 50, 1400

Andrew Smith, 150, 570, 5000, 125, 800

George Edwards 10, 30, 20, 25, 150

John Cummings, 40, 120, 800, 50, 350

William Seaberry, 15, 25, 200, 15, 225

John C. Love, 100, 380, 4000, 150, 60

David T. Freeman, 50, 70, 900, 100, 500

Henry Ritter, 30, 20, 300, 50, 350

Harman Strickland, 60, 140, 800, 100, 400

William Rowan, 40, 260, 1500, 50, 250

Robert Houghton, 450, 1050, 7500, 500, 2400

Hestar Patterson, 30, 90, 300, 25, 230

Hugh P. Mitchell, 100, 420, 2600, 150, 625

John M. Walker, 30, 50, 00, 45, 320

William Wright, 20, 260, 500, 35, 550

John W. Davidson, 30, 100, 700, 30, 300

John Hare, 14, 115, 700, 50, 350

William E. Cooper, 9, 71, 700, 25, 200

Durham G. Saunders, 100, 300, 1000, 100, 600

Robert A. Bradham, 40, 960, 8000, 50, 500

William Speight, 150, 250, 1600, 150, 650

Alexander Croom, 800, 800, 10000, 1000, 3000

Archibald McPhottis, 60, 100, 800, 75, 800

Donald Nicholson, 500, 1000, 4500, 200, 1700

Richard Hagood, 200, 200, 2500, 200, 520

Ann McIntosh, 40, 40, 400, 75, 150

Augustus Johnson, 500, 900, 8500, 350, 1100

Alexander Kennedy, 30, 30, 600, 50, 550

Joseph Seabrook, 300, 260, 4500, 250, 1200

Jeptha Gregory, 125, 295, 2000, 150, 750

Francis Farmer, 15, 155, 800, 20, 140

Arthur Truelock, 18, 22, 150, 15, 112

Edwin Thigpin, 25, 15, 500, 25, 450

William H. Armistead, 250, 673, 5000, 175, 1000

Charles Blount, 30, 100, 1000, 50, 350

Herndon Fykes, 20, 180, 800, 25, 110

James Pittman, 70, 100, 1000, 150, 400

Stephen Chason, 6, 100, 500, 25, 250

Stephen Browning, 200, 360, 3000, 200, 800

Robert Tucker, 15, 150, 250, 25, 140

Alfred Sheppard, 130, 150, 1500, 200, 1000

John H. Harrison, 60, 100, 250, 45, 200

Joseph Chandler, 300, 600, 4500, 250, 850

Seth Hadapon, 150, 120, 540, 175, 800

James Thomas, 350, 1250, 10000, 250, 2000

Spencer Short, 80, 100, 1000, 125, 350

John Smith, 80, 80, 500, 50, 1100

Richard King, 40, 60, 500, 50, 525

Asa Barber, 75, 635, 4000, 200, 550

Lewis B. Gregory, 100, 340, 3000, 180, 450

John W. Jones, 180, 300, 1500, 125, 650

Robert Marlow, 60, 100, 1000, 100, 450

William Egerton, 150, 50, 2000, 200, 750

George W. Floyd, 15, 65, 500, 50, 1000

James M. Ellis, 43, 257, 500, 50, 325

James Floyd, 25, 100, 500, 25, 700

Alexander Campbell, 200, 500, 4000, 500, 1200

Jefferson Davis, 150, 230, 2500, 500, 900

Daniel Bradwell, 300, 1300, 4000, 600, 1750

Stokely Sadbury, 25, 55, 400, 125, 600

Bennett Battle, 250, 150, 1500, 300, 900

Benjamin Womack, 40, 60, 300, 50, 900

James Thomas, 20, 60, 200, 50, 109

Abner Linn, 80, -, 300, 125, 725

William M. Linn, 12, 148, 1500, 125, 400

Walter Robinson, 25, 55, 400, 75, 450

Henry Chason, 17, 98, 1000, 50, 250

Jonathan Ellis, 70, 330, 1000, 50, 400

William C. Sivert, 120, 130, 1500, 225, 700

James K. Oates, 14, 36, 150, 35, 150

David D. Connell, 30, 90, 250, 40, 140

Thomas Connell, 20, 60, 00, 35, 500

Daniel Browning, 90, 510, 4000, 200, 700

William Gregory, 35, 95, 525, 85, 225

James G. Allen, 15, 65, 200, 50, 75

Henry M. Jones, 130, 390, 2500, 250, 650

Archibald Cambbell, 50, 250, 1500, 175, 1000

Newett (Hewett) Gilchrist, 150, 350, 2500, 180, 900

James M. Elliott, 200, 400, 3000, 200, 1000

Laurence Lines, 150, 500, 5000, 200, 1200

Joseph Ostvold, 25, 55, 200, 50, 150

Weem Arnold, 30, 170, 1000, 100, 400

John M. Woster, 30, 162, 1500, 125, 850

Thomas Edwards, 60, 20, 300, 75, 700

Izdaliah Wood, 100, 80, 400, 100, 500

Joshua Davis, 200, 680, 6000, 300, 850

George Martin, 40, 80, 600, 125, 350

Washington Wilkes, 45, 90, 650, 85, 375

John Smith, 20, 20, 200, 40, 250

Henry Perkins, 40, 80, 600, 75, 330

Jesse McCall, 400, 350, 10000, 1000, 1400

Angus Finlayson, 100, 140, 1200, 200, 800

John F. Thomas, 26, 54, 150, 75, 310

Fardine Worthington, 120, 240, 1500, 100, 550

Stephen Dixon, 75, 85, 1000, 85, 500

William Edwards, 150, 250, 3000, 150, 1900

Alexander Love, 100, 140, 2000, 150, 900

J. Hugh Mouchet, 60, 940, 2000, 125, 520

Daniel Johnson, 30, 10, 100, 15, 650

Alexander Johnson, 15, 65, 100, 50, 500

Samuel Patterson, 50, 110, 200, 75, 600

Archibald Buie, 17, 53, 100, 35, 127

Jesse Harper, 35, 115, 500, 125, 250

John Wabbleter, 20, 280, 1000, 100, 260

Stephen Leely, 40, 280, 1000, 200, 500

John T. Smith, 50, 270, 1500, 200, 1150

Randall Johnson, 100, 200, 1500, 150, 1000

_uchins Sheppard, 25, 55, 600, 75, 400

Joseph Sheppard, 18, 22, 300, 50, 250

Jeremiah Nepel, 20, 20, 200, 50, 220

William Sheppard, 20, 20, 200, 75, 300

Jacob Sheppard, 50, 110, 480, 80, 450

John McLene, 15, 35, 400, 45, 380

Richard McComb, 35, 70, 300, 60, 400

Mary Hare, 30, 50, 500, 75, 300

Benjamin Bateman, 80, 120, 1000, 150, 760

Richard Taft, 20, 20, 100, 40, 320

William Thomas, 30, 50, 800, 75, 550

Andrew Arnold, 12, 28, 100, 40, 200

John Graham, 30, 130, 500, 70, 450

Jonas Mathews, 20, 20, 100, 40, 250

Horace Vann, 25, 55, 250, 75, 250

Henry Lamb, 25, 55, 300, 100, 440

John Song (Sones), 32, 128, 800, 125, 425

Abraham Mathews, 20, 20, 150, 75, 450

John McRae, 35, 265, 2500, 150, 340

Charles Gregory, 100, 640, 10000, 300, 1040

Mary McGill, 25, 135, 500, 100, 270

Wright Beckwith, 30, 50, 400, 75, 300

James Ferguson, 30, 120, 300, 80, 320

Harvey Boykin, 19, 100, 400, 70, 440

John L. Smith, 20, 100, 400, 125, 350

John Gripard, 35, 605, 8000, 350, 850

Reuben Scarborough, 125, 275, 2000, 200, 750

John Luton, 175, 525, 5000, 400, 950

Cornelius English, 50, 650, 8000, 350, 850

Jonah English, 12, 138, 1000, 70, 150

Jackson Johnson, 25, 515, 4000, 200, 1050

David Rhymes, 17, 273, 1000, 100, 250

Edward Sykes, 25, 185, 1500, 85, 460

Edmund Harrison, 20, 60, 800, 75, 380

Jackson Roberts, 20, 80, 300, 75, 550

Joseph Lane (Sane), 20, 20, 400, 50, 160

Josiah Joucer (Jouces), 15, 25, 200, 25, 265

John Carver, 25, 400, 1000, 70, 910

Lyttlebury Parker, 40, 200, 1000, 125, 800

William Dixon, 22, 18, 400, 45, 400

William Wilson, 44, 36, 800, 90, 570

William McClenden, 20, 980, 4000, 75, 470

Francis Franklin, 20, 67, 500, 45, 400

Allen Angling, 15, 50, 670, 50, 280

Joseph Blanchard, 40, 40, 800, 75, 450

John Collyer, 17, 23, 300, 25, 200

Daniel Wilder, 40, 300, 1700, 50, 350

James Storm, 30, 270, 1000, 85, 400

Lewis Gregory, 60, 940, 7000, 400, 4500

Thomas Nixon, 25, 500, 5000, 200, 700

James (Jane) Byrd, 20, 30, 500, 60, 300

Jesse Dukes, 20, 100, 1500, 65, 400

Duncan Johnson, 20, 20, 300, 40, 200

Frances Byrd, 10, 200, 1000, 65, 330

Wyatt Hernden, 20, 70, 900, 45, 310

Hardy Glenn, 12, 150, 500, 35, 300

William S. Larkin, 25, 175, 1000, 80, 800

Jacob Wallace, 10, 30, 400, 25, 160

Thomas Stedstek, 20, 100, 500, 75, 400

John Chason, 60, 640, 2000, 125, 1200

Elijah Stanford, 20, 100, 600, 45, 400

John Pig, 35, 1465, 5000, 85, 1750

Edward Landingham, 28, 175, 700, 75, 450

Archibald Martin, 15, 90, 500, 45, 400

Simon Goff, 40, 100, 650, 50, 500

Joseph Chason, 19, 301, 400, 35, 580

Benjamin Holly, 15, 65, 300, 40, 235

Wiley Parish, 22, 18, 150, 25, 200

Daniel Statameyer, 50, 70, 600, 125, 520

Levi Butler, 27, 13, 100, 40, 175

John Strickland, 20, 100, 500, 25, 220

Edward Crosby, 50, 200, 500, 25, 400

Wm. Bryant, 40, 10, 350, 75, 350

Wasers Bryant, 35, 105, 350, 50, 230

Jones Huckaby, 18, 22, 150, 25, 250

Beryholl Gandy, 25, 95, 400, 120, 450

William Cash, 35, 400, 2000, 75, 650

Seaborn Pelt, 15, 25, 200, 20, 150

Isham Strickland, 18, 22, 150, 45, 250

Beniah Boatwright, 23, 17, 150, 30, 225

Thomas Benbry, 40, 10, 600, 75, 440

Andrew Patterson, 15, 65, 300, 25, 250

John F. Brown, 35, 145, 900, 125, 400

Washington Braddock, 90, 210, 1500, 250, 850

James Nims, 350, 1250, 4000, 300, 1700

James M. Stafford, 25, 15, 200, 65, 300

Hardy Todd, 40, 200, 1600, 80, 400

Benjamin Eubanks, 19, 21, 200, 50, 350

Richard Gatlin, 25, 53, 400, 125, 400

Elias Rudd, 30, 90, 600, 90, 400

Samuel Rudd, 15, 35, 180, 40, 100

Pharoah Cross, 36, 44, 250, 75, 320

John Spinx, 20, 20, 200, 40, 175

Sarah Warren, 30, 90, 450, 80, 350

Presley Tharp, 25, 15, 200, 60, 300

Jane Gibson, 800, 1200, 15000, 1000, 2800

John Edenfield, 50, 138, 500, 125, 375

Henry Toler, 45, 35, 200, 40, 250

Henry Smith, 40, 130, 4000, 125, 420

Robert M. Witherspoon, 150, 150, 1200, 200, 700

Jesse Woods, 100, 140, 2000, 120, 550

William Seely, 40, 80, 250, 95, 425

Matthew McLilly, 25, 15, 1000, 70, 250

Marcellus Morgan, 70, 220, 2000, 175, 850

Jesse Gregory, 300, 441, 6000, 1000, 1300

William Dean, 55, 25, 300, 75, 325

William Williams, 75, 275, 2000, 125, 224

Owen Anders, 60, 340, 2000, 125, 625

Alexander Anders, 50, 110, 1000, 150, 600

John Smith, 34, 45, 300, 100, 400

Obadiah Michaux, 150, 250, 4000, 250, 1000

William G. Mayton, 75, 250, 2000, 145, 600

Starkey J. Cox, 45, 120, 600, 75, 400

John McElvey, 100, 340, 2000, 150, 1000

Lawson McElvio, 200, 400, 6000, 250, 1100

Solomon Owens, 75, 200, 1000, 125, 500

William H. Forester, 40, 80, 250, 80, 225

Unity Bryan, 40, 40, 175, 75, 250

Green McRae, 50, 30, 500, 125, 750

Redding Blount, 60, 40, 500, 125, 400

Edwin Blount, 40, 40, 350, 75, 340

John Spinx, 200, 270, 1500, 200, 700

Mary A. Strickland, 75, 245, 1000, 125, 650

George Fleming, 190, 210, 1700, 125, 600

Rebecca Davis, 50, 110, 300, 75, 330

Griffin Ball, 35, 45, 150, 50, 350

Joshua Stokes, 30, 50, 150, 125, 350

Asa Townsend, 25, 55, 150, 75, 375

Griffin Fletcher, 80, 80, 1500, 225, 670

Solomon Nelson, 40, 40, 200, 60, 350

David Alderman, 75, 230, 1000, 150, 950

Jesse Bryan, 60, 80, 700, 100, 670

Stephen Nelson, 40, 40, 240, 75, 320

David Mills, 80, 80, 800, 225, 1110

Cleatus Mills, 25, 45, 400, 75, 420

William Lott, 25, 15, 150, 80, 340

John G. Gornel, 20, 20, 500, 60, 250
John Whidder (Whedden), 30, 50, 600, 85, 370
Nelson Thompson, 40, 20, 300, 45, 200
Theophilus Thompson, 35, 45, 400, 150, 350
James Holly, 25, 55, 400, 75, 215
David Grey, 35, 45, 500, 120, 400
Simeon Vickers, 60, 60, 500, 140, 425
John G. Base (Bose), 40, 40, 300, 100, 335
Brison Vickers, 35, 45, 450, 75, 370
Jacob Dykes, 100, 100, 1000, 250, 650
Dixon Dykes, 40, 40, 400, 45, 150
John W. Mann, 125, 524, 4000, 250, 750
William Bolin, 35, 5, 150, 60, 200
Walton Lott, 100, 100, 800, 125, 670
James Barfield, 50, 133, 1000, 150, 300
William Ferrill, 20, 20, 200, 80, 230
George W. Kemp, 200, 300, 4000, 300, 400
Abraham Hand, 50, 30, 200, 50, 270
John Mann, 60, 50, 450, 125, 370
John Riggins, 70, 110, 1000, 150, 550
Joseph Dykes, 90, 110, 900, 200, 650
Henry Womack, 70, 80, 600, 140, 400
Levi Shelfer, 40, 40, 300, 80, 235
Alexander McKenzie, 120, 200, 2000, 300, 850
Alfred Vickers, 35, 45, 250, 90, 350
Uriah Vickers, 50, 30, 300, 125, 325
Nicholas Robinson, 100, 540, 2000, 3 00, 150
Walton Robinson, 50, 250, 1800, 150, 350
Gerard Barber, 60, 60, 500, 100, 430
William Barber Sr., 200, 440, 2000, 350, 1100

William Barber Jr., 70, 50, 800, 150, 520
Jordan Barber, 60, 60, 800, 125, 400
David Long (Lang), 100, 200, 1500, 250, 950
Margaret Henry, 125, 175, 1000, 250, 650
George Henry, 50, 30, 400, 75, 350
Edmund Shedo, 60, 50, 500, 75, 350
Walter Weech, 30, 50, 200, 50, 400
Bryan Vickers, 35, 125, 600, 125, 425
Amos Newton, 30, 50, 200, 125, 300
Ezekiel Vickers, 50, 30, 300, 100, 400
Jordon Dykes, 60, 20, 300, 175, 420
James Grey, 40, 40, 400, 125, 400
John Johnson, 60, 30, 500, 200, 400
Bennet Poppel, 50, 30, 400, 125, 470
Pharoah Poppel, 25, 55, 300, 25, 200
Appling Wishan, 30, 50, 350, 140, 350
Sarah Grice, 60, 40, 500, 150, 400
Raphael Butler, 50, 25, 400, 90, 165
Abner Chester, 80, 160, 500, 170, 530
Winnifred Spence, 60, 60, 600, 150, 530
George W. Finkley, 45, 85, 300, 100, 280
John Bradley, 50, 40, 450, 140, 350
Andrew Miller, 25, 55, 400, 100, 250
Nathan Shelfer, 100, 500, 2000, 200, 900
John Grey, 55, 25, 400, 125, 275
Thomas Collins, 25, 55, 250, 130, 250
Isaac King, 50, 70, 600, 150, 630
Lawrence Joiner, 60, 30, 450, 125, 350
Samuel Mashburn, 50, 30, 300, 100, 550
William McKenzie, 30, 50, 250, 150, 360
William Lott, 40, 40, 300, 75, 350
Nathaniel Lott, 35, 50, 250, 125, 335

Alley Wilson, 70, 80, 500, 200, 475

John Womack, 55, 25, 400, 125, 350

Zabud Fletcher, 200, 500, 3500, 300, 950

William Ball, 50, 30, 400, 100, 370

Willis Hudnel, 300, 160, 5000, 350, 1100

Theoditus Hudnel, 60, 44, 1000, 125, 280

William Stafford, 250, 750, 4000, 300, 1400

Thomas Scott, 160, 440, 3000, 200, 750

Chrice Hall, 400, 700, 6000, 350, 1900

William Gibson, 600, 1400, 8000, 300, 1700

Catharine Woodburn, 350, 290, 5000, 150, 1160

Patrick Thomas, 60, 80, 700, 175, 375

Thomas Mills, 80, 200, 120, 500, 550

Augustus Lanier, 800, 500, 10000, 300, 3500

William Johnson, 500, 700, 5000, 350, 1600

Robert H. Harrison, 300, 1533, 17000, 100, 1460

Jackson McCoy, 40, 40, 300, 400, 335

William Rodgers, 350, 650, 7000, 65, 1250

Drew Glover, 35, 5, 140, 70, 175

James Mashburn, 40, 40, 250, 125, 350

Easton Whidden, 28, 12, 178, 125, 40

John McMillan, 300, 700, 8000, 500, 1500

Wilson Thurman, 60, 60, 450, 125, 400

Daniel McDaniel, 50, 30, 375, 150, 670

Duncan McPhaul, 35, 45, 300, 125, 400

Creed Boykin, 25, 25, 200, 70, 215

Sarah Smith, 60, 60, 500, 140, 500

Alexander McPhaul, 25, 55, 250, 95, 200

George W. Sunday, 40, 40, 300, 125, 400

Joshua Johnson, 60, 128, 700, 150, 530

Robert Witherspoon, 30, 70, 500, 150, 950

Joseph O'neil, 80, 160, 1000, 200, 1000

Mark Thomas, 40, 40, 200, 75, 160

Alexander Gregory, 80, 120, 1000, 150, 600

James G. Owens, 25, 195, 1300, 150, 375

Edward Dixon, 45, 35, 400, 175, 550

Mary Mitchell, 40, 80, 500, 150, 500

Thomas Goza, 60, 100, 1000, 150, 580

John Renew, 40, 40, 250, 75, 195

Oliver E. McKeown, 35, 35, 300, 100, 400

Mary Gerow, 35, 45, 200, 125, 260

John Saddler, 40, 40, 320, 140, 350

Rachael Mathison, 50, 150, 1000, 150, 400

Amily Gregory, 50, 110, 1500, 125, 620

Allen McKenzie, 100, 140, 1500, 150, 660

Jacob Wilder, 50, 70, 400, 95, 370

Nehemiah Mercer, 25, 115, 400, 125, 340

Isham Kain, 45, 35, 400, 140, 365

Joseph Muscolina, 40, 120, 450, 51, 150

Harriet Carnochin, 400, 560, 5000, 250, 1900

John Carnochin, 140, 500, 3000, 175, 300

Gideon Hawkins, 50, 190, 1500, 100, 590

Joseph Sylvester, 300, 430, 6000, 350, 1600

Hamilton County, Florida
1850 Agricultural Census

The University of North Carolina at Chapel Hill filmed the 1850 agricultural census for Hamilton County from originals at the Florida State University under a grant from the National Science Foundation in 1963.

Columns 1, 2, 3, 4, 5, and 13 represent the following information on the census:
1. Name of Owner, Agent or Manager of Farm
2. Acres of Improved Land
3. Acres of Unimproved Land
4. Cash Value of the Farm
5. Value of Farming Implements and Machinery
13. Value of Livestock

Sarah Taylor, 25, -, 150, 50, 350
James Oglesbay, 50, -, 300, 30, 360
Daniel K. Shaw, 40, -, 200, 20, 175
John G. Slade, 15, -, 200, 75, 360
Joseph Wilson, 40, -, 300, 100, 520
James R. Oliver, 40, -, 400, 55, 442
Lemuel Taylor, 16, -, 150, 5, 125
John Williams, 50, -, 600, 60, 260
Matthew McCullers, 20, -, 100, 5, 305
William S. Bryan, 25, -, 100, 35, 190
Dempsey D. Crews, 30, -, 150, 15, 435
Lewis H. Bryan, 20, -, 200, 20, 360
William W. Williams, 30, -, 150, 50, 366
Joseph A. Ellis, 40, -, 225, 50, 100
John Bryan, 25, -, 200, 25, 930
Philemon Bryan, 60, -, 200, 20,700
Milton J. Bryan, 80, -, 420, 130, 1500
Haynes Bryan, 30, -, 50, 8, 200
Jehu Blunt, 75, -, 600, 120, 340
Crawford Parish, 200, -, 1000, 200, 760
Pernell Cason, 30, -, 250, 50, 275
James Geiger, 16, -, 100, 10, 130
Cullen W. Cason, 30, -, 250, 35, 200
Stephen Locke, 50, -, 250, 15, 240
Abigail Cason, 25, -, 150, 5, 330

William R. Penington, 40, -, 400, 90, 400
Edward Lee, 22, -, 200, 8, 233
Joshua H. Roberts, 35, 30, 350, 50, 240
Levi Lee, 60, -, 500, 30, 400
John R. Garrett, 25, -, 200, 35, 284
James L. Ross, 75, 34, 1000, 55, 305
Zilpha Hogans, 35, 40, 400, 100, 780
John Lee, 82, -, 1000, 325, 1390
James McDonald, 45, 315, 1000, 50, 660
James D. Prevatt, 40, 200, 1400, 50, 400
Willey Lee, 35, 10, 500, 80, 460
Henry E. Parriance, 200, 800, 4000, 300, 1000
Bryant Sheffield, 80, 160, -, 300, 920
William M. Reed, 40, 360, 2000, 60, 360
Francis J. Broward, 55, 35, 50, 75, 530
Benjamin Jackson, 55, 65, 400, 60, 190
Elizabeth Stapleton, 60, 340, 2000, 60, 305
Thomas W. Smith, 45, -, 200, 30, 148
Abram B. Smith, 35, -, -, 40, 230
Samuel B. Foster, 30, 10, 150, 35, 295

William H. Dowling, 25, -, 300, 60, 100

Barnabas Chesire, 20, -, 200, 15, 230

Amos Cheshire, 25, -, 135, 5, 35

William Herren, 50, 20, 200, 35, 260

John L. Jerry, 240, 600, 6000, 300, 890

William Cheshire, 100, -, 300, 50, 600

Joseph Bellott, -, -, -, 50, 200

James Cheshire, 20, -, 150, 45, 290

William Yearty, 20, -, 20, 30, 246

Elihu Oglesbay, 35, 165, 3000, 60, 300

Mary Holmes, 20, -, 100, 25, 280

Jeremiah Johns, 65, 15, 500, 10, 860

Hezekiah Johns, 23, -, 140, 10, 200

Francis J. Ross, 220, 580, 2000, 500, 2160

Tapley A. Tillis, -, -, -, 40, 285

John G. Smith, 100, 87, 800, 150, 680

Absalom Smith, 15, -, 200, 10, 280

Edmund M. Smith, 20, -, 100, 3, 300

Bunyan Mathis, 18, -, 125, 10, 110

George B. Smith, -, -, -, 100, 300

Edmund Thompson, 10, -, 65, 10, 46

Robert Ivey, 25, 25, 350, 10, 120

James R. Turner, 70, 40, 700, 100, 486

John Turner, 20, 20, 200, 50, 160

Adam S. Goodbread, 200, 340, 4000, 450, 1404

Thomas Bryan, 40, -, 400, 50, 433

Thomas Beal, 40, -, 400, 35, 247

James Brewer, 25,-, 400, 80, 273

James H. Prevatt, 25, -, 150, 15, 270

John R. Cheshire, 28, -, 150, 25, 265

Solomon Robson, 65, 135, 1500, 80, 345

Jonathan K. Prevatt, 25, 55, 500, 20, 300

William J. D. Prevatt, -, -, -, 8, 225

Turner Jackson, 100,-, 500, 225, 445

Thomas Altman, 30, -, 400, 15, 350

James R. Smith, 40, -, 200, 35, 245

Jesse A. Swilley, 60, 100, 900, 20, 175

Arther Smith, 15, -, 200, 10, 190

Lewis L. Fortic, 40, -, 400, 25, 105

Guilford Register, 100, 300, 800, 50, 2200

William Dougharty, 70, 10, 1000, 75, 700

Henry Penington, 50, 270, 1500, 120, 1450

Abraham Geiger, 20, 100, 700, 6, 90

Hiram Sanders, 40, 70, 300, 37, 560

William Stephens, 20, -, 100, 5, 100

Joshua Stephens, 40, -, 300, 40, 630

Samuel Cox, 50, -, 300, 30, 550

Benjamin Rowlins, 80, 120, 80, 90, 548

William Yeats, 75, -, 400, 70, 860

Samuel Knight, 15,-, 60, 40, 130

Henry J. Stewart, -, -, -, -, 425

Charles Hendry, 40, -, 350, 10, 260

Absalom S. Smith, 100, 340, 500, 230, 1100

Hezekiah Harbuck, 25, 17, 1200, 80, 130

William Millican, 140, 100, 1000, 150, 700

John Roberts, 200, 120, 1500, 100, 700

James H. Stephens, 75, -, 500, 40, 500

Seaborn Lastinger, 60, -, 300, 80, 490

William Tyre, 20, -, 150, 30, 120

Bryant Burnett, -, -, -, 10, 300

William S. Fergerson, -, -, -, 10, 150

John E. Tuten, 156, 120, 2500, 200, 700

Joseph E. Law, 100, 100, 1500, 70, 523

William M. Hunter, 15, 24, 100, 125, 620

William Ponchier (Pouchier), 35, 125, 800, 30, 204

Millington M. Smith, 30, 45, 250, 35, 225

Abraham Knight, 14, -, 100, 10, 215
Thomas Knight, 25, -, 200, 60, 400
Redden J. Tuten, 35, -, 250, 30, 260
John Pierson, 45, -, 300, 40, 312
Thomas J. Stewart, 40, -, 800, 50, 400
Willis W. Burke, 160, 593, 2000, 210, 1790
Thomas T. Wright, 90, -, 400, 200, 800
John J. Underwood, 35, 47, 700, 10, 735
Thomas T. Brooke, 50, -, 300, 50, 480
Daniel Bell, 120, 1180, 2000, 200, 1500
Israel _. Stewart, 200, 700, 1800, 150, 886
Nathaniel P. Marion, 400, 300, 3000, 300, 1035
Stewart A. Mitchel, 30, 390, 800, 211, 552
Robert Hendry, 60, 20, 500, 130, 264
William Parish, -, -, -, 40, 60
James M. Claridy, 55, 185, 1000, 10, 300
Gabriel Hall, 15,-, 150, 20, 520
James McCradie, 20, 140, 400, 20, 288
George Mikell, 38, -, 250, 25, 265
John Ruskin, -, -, -, 20, 100
Thomas Jordan, 70, 120, 500, 15, 175
James B. Gore, 20, -, 130, 85, 128
James Duncan, 210, 150, 1000, 100, 665
Nehemiah Hall, 25, 70, 500, 15, 229
William M. Hunter, 75, -, 250, 100, 735
William Day, 40, 120, 800, 40, 240
John Hall, 30, 10, 500, 80, 180
James Mims, 40, 160, 800, 30, 1100
George S. Jennings, 55, 25, 700, 175, 440
George Jennings, 200, -, 1500, 225, 425

John S. Sharpe, 35, -, 375, 25, 400
William T. Jennings, 60, 60, 600, 25, 300
Elijah Deas, 18, -, 100, 10, 105
Luther P. Wrede, 60, 130, 600, 30, 300
Daniel High, 20, -, -, 5, 100
John Zipperer Jr., 25, 80, 100, 15, 370
Josiah T. Baisden, 75, 285, 1000, 200, 640
William Bassett, 18, 62, 100, 25, 134
Lewis B. Towler, 14, -, 100, 5, 130
John Padgett, 30, 10, 400, 35, 212
Andrew J. Leigh, 20, -, 100, 30, 360
William Robuck, 50, 110, 900, 35, 440
William R. Towles(Towler), 22, 148, 400, 15, 327
Armand Lefils, 18, -, 130, 30, 130
James Hendry, 68, 135, 400, 50, 334
Jeremiah Smith, 40, -, 150, 5, 174
Handen Cheshire, 30, 20, 500, 40, 447
George Johnson, 57, -, 400, 60, 382
Gibson Blalock, 65, 58, 500, 125, 416
Elijah Bersley, 28, -, 300, 30, 320
Elzy B. Lealman, 25, -, 150, 30, 135
John Sasser, 50, -, 300, 20, 830
Terese Taylor, 20, -, 200, 5, 225
Joseph Deas, 20, -, 175, 5, 125
Leonard Deas Sr., 118, -, 600, 261, 860
Solomon Hedgecock, 20, -, 1000, 25, 284
John M. Zipperer, 25, -, 200, 10, 150
John Platt, 17, -, 100, 35, 125
James Farnell, 40, -, 200, 8, 191
Allen G. Johnson, 140, 1100, 1500, 350, 1200
Isaac Ogden, 26, 20, 100, 22, 121
James S. Bell, 120, 1400, 7000, 300, 800
Richard Herndon, 75, -, 500, 40, 200
Jane J. Polhill, 70, 120, 800, 75, 430

James T. Stewart, 20, 60, 500, 40, 390

Mathew M. Deas, 150, 30, 900, 200, 864

James L. Anderson, 50, 150, 1200, 200, 730

John S. Deas, 35,-, 200, 25, 250

Henry Hunter, 30, -, 60, 10, 180

Lewis M. Deas, 25, -, 250, 66, 270

William Moody, 50, 30, 600, 188, 3560

Leonard M. Deas, 30, -, 250, 35, 155

Aaron M. Deas, 50, 30, 800, 35, 282

David B. Norman, 12, -, 200, 6, 130

Renatus Downing, 14, -, 200, 50, 130

Jarad Westbury, 20, -, -, 15, 75

Jacob N. Driggers, 20, -, 200, 15, 152

Martha Robuck, 60, 250, 1000, 140, 420

Jeremiah B. Smith, 60, -, 1400, 40, 325

Henry M. Stephens, 200, 420, 1800, 225, 1350

James Burnum, 10, 260, 2000, 100, 550

Elizabeth Kersh, 17, 23, 200, 2, 100

Samuel R. Johnson, 75, 120, 800, 165, 233

David Moat, 40, -, 1000, 75, 687

Stephen Kersh, 20, -, 300, 35, 350

James W. Gill, 25, -, 250, 50, 238

William Dempsey, 28, 52, 400, 45, 336

Handerson Graham, 25, 215, 500, 15, 200

Daniel Willis, 35, -, 250, 35, 250

Edwin F. Wrede, 25, -, 200, 25, 247

Solomon B. Smith, 130, 30, 1000, 100, 575

James Hall, 26, -, 150, 68, 520

Hillsborough County, Florida
1850 Agricultural Census

The University of North Carolina at Chapel Hill filmed the 1850 agricultural census for Hillsborough County from originals at the Florida State University under a grant from the National Science Foundation in 1963.

Columns 1, 2, 3, 4, 5, and 13 represent the following information on the census:
1. Name of Owner, Agent or Manager of Farm
2. Acres of Improved Land
3. Acres of Unimproved Land
4. Cash Value of the Farm
5. Value of Farming Implements and Machinery
13. Value of Livestock

Elias J. Hart, 59, 300, 3600, 100, 550
Samuel H. Stevenson, 10, 190, 900, 50, 375
William Jackson, 2, -, 90, 29, 550
Jesse Carlisle, 14, 320, 1000, 45, 1200
Joseph Daniels, 3, -, 50, 15, 310
John Whitehurst, 16, 144, 800, 20, 675
Samuel Houlair, 6, 74, 300, 50, 214
James McNeil, 6, -, 170, 65, 700
William Brooker, 3, -, 100, 25, 54
Asa McClenden, 20, 150, 670, 69, 1080
S. A. Branch, 13, -, 100, 10, 40
Francis Mathews, 9, 40, 300, 15, 70
John Green, 10, -, 200, 30, 400
Richard V. Buffain, 3, -, 50, 5, 740
William Miley, 10, 30, 200, 35, 380
George Brace, 4, -, 200, 5, 3
Lamar Stevens, 5, -, 300, 25, 550
James M. Cooker, 4, -, 135, 11, 174
Daniel Giller, 12,-, 208, 25, 1146
John Carney, 20, -, 695, 43, 743
Gideon Hagar, 20, -, 710, 41, 1274
Robert Gamble Jr., 330, 960, 23500, 25000, 2250
William P. Craig, 500, 1060, 20000, 3000, 1200
Franklin Branch, 25, 975, 2500, 220, 550

Ezekiel Glazier, 10, 190, 218, 5, 248
H__ S. Clark, 10, 75, 1203, 45, 425
Josiah Gates, 43, 217, 14000, 2000, 1340
German H. Wyatt, 6, -, 250, 11, 5000
John Addison, 10, -, 250, 10, 550
William Addison, 3, -, 100, 5, 350
Joseph A. Braden, 300, 600, 50000, 10000 1400
Elaxander F. Forester, 6, 74, 119, 5, 65
Edmund Whitaker, 10, 140, 930, 58, 225
William Whitiker, 20, 180, 2500, 20, 1000
William Weeks, 70, 140, 700, 100, 600
James Rains, 15, 145, 250, 150, 3045
Thomas Mitchell, 20, 140, 800, 150, 700
Thomas P. Kennedy, 40, 300, 1600, 100, 250
William F. Lockwood, 8, 193, 700, 25, 350
Simeon Turner, 3, 160, 300, 5, 50
Thomas Cowart, 30, 792, 640, 80, 1600
James Stevens, 40, 600, 1600, 35, 1500
Jacob H. Miller, 8, 172, 200, 5, 160

Elisabeth A. Barry, 3, 160, 200, 5 , 130

Wm. H. McDonald, 25, 135, 200, 30, 228

William Cooley, 50, 640, 3000, 60, 400

William T. Brown, 40, 120, 1000, 200, 300

Ann M. Roberts, 105, 475, 4000, 400, 800

John W. Roberts, 8, 72, 50, 100, 137

Jesse Carter, 2, 400, 1000, 12, 900

Thos. Macquire, 300, -, 500, 20, 225

Jane T. Magbee, 20, 203, 1200, 25, 80

Samson Forester, 20, 20, 300, 16, 1475

Aadam Fatis, 4, 20, 100, 5, 470

Louis Bell, 10, 150, 500, 5, -

J. B. Burisell, 65, 415, 2400, 75, 500

Daniel P. Myers, 4, 40, 40, 50, 300

Benjamin Moody, 12, 148, 350, 20, 460

Christy Fribly, 80, 799, 300, -, 300

A. J. Snider, 2, 158, 1000, 5, 400

James White, 8, 172, 210, 10, -

Seth Howard, 6, 600, 160, 20, 300

Whitton William, 13, -, 120, 15, 320

William B. Hooker, 250, 1230, 7500, 250, 13358

William Parker, 12, 148, 600, 50, 1000

John C. White, 70, 70, 1000, 20, 90

James Oliver, 45, 200, 250, 200, 250

John Parker, 30, 130, 700, 25, 1200

John Pearse, 16, 150, 400, 20, 700

John Gallagher, !0, 150, 350, 40, 800

C. B. Sparkman, 20, -, 400, 25, 1600

Samuel Rogers, 17, 143, 600, 5, 320

Stephen Hollingsworth, 29, 51, 500, 60, 600

Simon L. Sparkman, 50, 270, 1000, 700, 3000

Willoughby Whitten, 15, 25, 300, 25, 1500

William Whitton, 10, 30, 300, 10, 500

Moeses Turner, 16, 200, 200, 15, 270

Samuel Knight, 45, 35, 2000, 100, 2000

John Futch, 45, 115, 250, 40, 2200

William Hancock, 40, 220, 964, 50, 8000

William R. Russer, 15, 200, 200, 20, 600

Henry H. Frier, 12, -, 210, 20, 1100

Richard A. Vickers, 6,-, 100, 10, 500

Jesse Knight, 12, 148, 300, 105, 700

Frederick Varn, 30, 50, 600, 100, 800

Peter Platt, 12, 28, 500, 112, 1650

John Skipper, 7, -, 120, 5, 375

D. J. W. Boney, 6, -, 125, 20, 400

Rebon Rolinson, 6, -, 250, 55, 950

Elly Whitton, 13, -, 200, 10, 550

Jacob Summerline, 40, 200, 1500, 200, 3620

Rigdon Brown Jr., 15, -, 100, 3, 55

George Hamilton, 12, -, 150, 50, 1500

Maxfield Whitton, 16, 21, 230, 30, 2500

Joseph Howell, 8, 22, 250, 55, 1700

Silas Maclendon, 35, 150, 400, 35, 2700

John Thomas, 20, 300, 350, 25, 900

John Rolinson, 15,-, 400, 25, 600

John Vickers, 8, -, 300, 25,-

William Wiggins, 25, 150, 450, 25, 610

John Maclendon, 20, 100, 1000, 25, 120

Isaac Cruise, 20, 140, 500, 25, 470

Thomas Summerall, 25, 200, 300, 23, 255

Joe Summerall, 20, 200, 312, 25, 450

John Mercer, 15, 200, 300, 25, 2500

David Summerall, 20, 10, 150, 25, 204

Joseph Underhill, 8, -, 152, 63, 1015

William A. Willingham, 20, 140, 50, 50, 750

Benjamin Hilliard, 10, 150, 100, 10, 300

George Tyson, 6, -, 100, 30, 200

Mary Hall, 15, -, 100, 30, 700

Benjamin Guy, 6, -, 250, 10, 54

Jane Whitton, 51, 110, 1888, 46, 634

James Lanier, 52, 108, 1120, 85, 1806

Alderman Carlton, 20, -, 325, 32, 1114

Levy Pearce, 26, 14, 634, 28, 780

John Pearce, 1, -, 213, 23, 776

John C. Oats, 10, -, 400, 20, 100

Holmes County, Florida
1850 Agricultural Census

The University of North Carolina at Chapel Hill filmed the 1850 agricultural census for Holmes County from originals at the Florida State University under a grant from the National Science Foundation in 1963.

Columns 1, 2, 3, 4, 5, and 13 represent the following information on the census:
1. Name of Owner, Agent or Manager of Farm
2. Acres of Improved Land
3. Acres of Unimproved Land
4. Cash Value of the Farm
5. Value of Farming Implements and Machinery
13. Value of Livestock

James Evans, 30, -, 100, 5, 150
Jarrot Miller, 25, -, 100, 5, 160
James Barnes, 70, -, 100, 12, 325
Benj. T. Hague, 40, -, 80, 8, 150
Harrison Hagen, 12, -, 40, 5, 90
Stephen Hagen, 74, 50, 80, 6, 286
Thomas Young, 40, -, 120, 5, 100
Benj. Pitts, 15, -, 80, 5, 75
John Chapman, 60, 20, 500, 20, 175
Eli Wright, 40, -, 235, 35, 170
Wilson Musgroves, 50, -, 100, 18, 300
James Pittman, 14, -, 100, 5, 200
John Burton, 25, -, 75, 160, 350
James Taylor, 50, -, 100, 15, 185
Zilpha Gauf, 20, -, 150, 25, 250
Abram Faircloth, 20, -, 100, 5, 140
William Albritton, 35,-, 400, 20, 170
Samuel Pace, 60, -, 200, 5, 225
Richard Boyd, 35, -, 200, 40, 225
Thomas Jones, 45, -, 200, 40, 320
Thomas Pittman, 25, -, 250, 21, 360
Clark Braxon, 60, 30, 300, 70, 1200
Dempsey Farnell, 35, -, 100, 25, 360
A. B. Turner, 150, -, 1000, 300, 850
A. Turner, 35, -, 100, 10, 100
Major H. Stanley, 50, -, 150, 20, 140
William Cherens, 18, -, 100, 6, 95
George Turner, 20, -, 75, 5, 135
Albert Parish, 40, -, 100, 50, 140
James Turner, 80, 143, 150, 50, 750

Solomon Slaughter, 16, -, 100, 5, 110
Henry Jones, 25, -, 80, 3, 115
James Redick, 44, -, 100, 40, 80
John C. Knight, 80, -, 300, 30, 210
Robert Knight, 25, -, 300, 10, 300
Stephen Turner, 150, -, 300, 50, 400
James Turner, 30, -, 50, 6, 110
John Morrison, 75, -, 150, 250, 925
Margaret Gillis, 75, -, 150, 250, 1500
John Crempler, 40, -, 120, 25, 100
William Turner, 30, -, 100, 30, 240
Daniel Fowler, 20, -, 50, 10, 105
John Cruthfield, 20, -, 60, 16, 120
Jackson Cannon, 20, -, 75, 25, 500
Alexander Parker, 30, -, 150, 30, 475
Robert Crawford, 30, -, 100, 35, 175
William J. Yerly, 20, -, 100, 15, 75
Moses Christman, 25, -, 150, 50, 105
Jane Wilkes, 25, -, 150, 20, 108
John Owens, 40, -, 150, 60, 800
John Willis, 14, -, 50, 13, 265
Jane Whittaker, 30, -, 175, 30, 450
Allen Register, 30, -, 100, 10, 205
Ezekiel Register, 30, -, 100, 20, 507
Peter Grant, 30, -, 200, 20,188
Jacob Cravy, 15,-, 50, 20, 118
James Yates, 25, -, 100, 25, 5430
Allen Gibson, 25,-, 150, 45, 1450
Owen Williams, 80, 120, 1000, 200, 2105

Martin Mayo, 25, -, 100, 10, 180
Alfred Mayo, 27, -, 70, 10, 100
Samuel Perkins, 20, -, 120, 12, 150
Luke Cooper, 25, -, 100, 25, 160
Thomas Andrews, 40, -, 200, 16, 1120
John Smith, 40, -, 300, 70, 160
John McKinzie, 25, -, 150, 30, 200
Michael Joseph, 60, -, 200, 75, 500
Abner Baker, 15, -, 150, 5, 200
Mathew Glisson, 25, -, 200, 75, 500
Thomas Braxon, 80, -, 300, 200, 1510
Daniel Anderson, 30, -, 100, 25, 160
Robert Braxon, 60, -, 357, 50, 440
James Spears, 20, -, 150, 16, 220
Green Norris, 25, -, 180, 20, 100
James Grush, 25, -, 180, 20, 140
James G. Godwin, 50, -, 100, 30, 130
Jane Yerby, 35, -, 100, 25, 210
Angus Gillis, 100, 380, 800, 50, 1125
James Powell, 35, -, 200, 10, 140
John G. Gillis, 30, -, 200, 5, 125
David Neal, 18, -, 150, 20, 120
Allen Morrison, 40, -, 300, 350, 341

Benjamin Andrews, 45, 8, 400, 81, 350
George Mayo, 20, -, 75, 30, 155
Jane Thomas, 20, -, 100, 5, 300
Alfred Mayo, 35, -, 200, 70, 380
Michael Oates, 50, -, 200, 150, 345
Thomas Hewett, 40, -, 100, 100, 640
Wiley Blount, 61, -, 90, 65, 240
Joseph Turner, 35, 20, 200, 5, 120
Timothy Green, 50,-, 100, 30, 234
Talbot Parish, 35, -, 50, 30, 150
Etheldred Hewett, 20, -, 150, 5, 160
Moses Hewett, 25, -, 150, 30, 210
Elizabeth Jones, 18, -, 75, 7, 150
Meredith Cruthfield, 8, -, 50, 10, 184
John Cooper, 20, -, 200, 30, 70
Washington Cooper, 18, -, 100, 100, 85
Barnabas Dicen (Dicer), 30, -, 300, 100, 890
Louis Miller, 25, -, 100, 10, 895
William Whitaker, 16, -, 200, 30, 260
Samuel Mayo, 15, -, 100, 10, 100

Jackson County, Florida
1850 Agricultural Census

The University of North Carolina at Chapel Hill filmed the 1850 agricultural census for Jackson County from originals at the Florida State University under a grant from the National Science Foundation in 1963.

Columns 1, 2, 3, 4, 5, and 13 represent the following information on the census:
1. Name of Owner, Agent or Manager of Farm
2. Acres of Improved Land
3. Acres of Unimproved Land
4. Cash Value of the Farm
5. Value of Farming Implements and Machinery
13. Value of Livestock

Rodolphus Garbett, 30, 10, 150, 25, 300

Walton L. Scurlock, 6, 274, 125, 10, 200

Harriet Pope, 50, 714, 200, 30, 86

Nelson Harrley, 100, 380, 1000, 200, 880

Jno. R. Anderson, 40, 260, 400, 50, 425

Wm. Ham, 20, 60, 100, -, 200

G. C. Bird, 100, -, 500, 50, 500

Jesse Yon, 74, -, 800, 25, 350

Taylor Caraway, 30, -, 90, 3, 120

J. W. F. Bird, 50, -, 200, 20, 150

Jesse Coe Sr., 1000, 2000, 10000, 280, 2000

Jno. M. Hansford, 30, 50, 100, 20, 350

Lenn Griffin, 120, -, 500, 95, 400

John Lamis (Lomis), 35, -, 200, 20, 195

B. H. Stone, 30, -, 120, 15, 125

John Bird, 100, 220, 300, 35, 224

James Yon, 75, -, 250, 20, 175

Wm. McDaniel, 65, -, 200, 20, 190

George Bullock, 80, -, 250, 30, 175

Thos. M. White, 50, 110, 5000, 500, 2000

Wade Keyes, 240, 230, 3000, 200, 668

Susan A. Ming, 300, 500, 1000, 75, -

Joseph Williford, 65, 95, 831, 20, 575

N. O. S. Staley, 100, 140, 600, 100, 450

Sears Bryan, 80, -, 1500, 20, -

Henry J. Dawson, 40, 40, 500, 25, 225

Robert S. Dickson, 160, 800, 1200, 40, 230

Geo. W. Robinson, 140, 260, 700, 20, 4610

Hugh C. Davis, 50, -, 100, 15, 150

Ashby J. Davis, 20, 40, 150, 10, 175

Wm. M. Lord, 30, -, 150, 15, 90

A. F. Sellers, 18, 180, 200, 15, 320

Benj. Hewett, 12, -, 120, 10, 110

John F. Sellers, 20, 70, 125, 15, 380

James Cloud, 16, 70, 130, 15, 250

J. J. Peacock, 25, 75, 200, 13, 175

William Yarborough, 70, -, 400, 75, 375

Iohn Cloud, 16, 64, 800, 5, 65

Wm. L. Corbett, 10, 70, 100, 7, 170

Thos. Hare, 8, 115, 80, 10, 950

William Hare, 12, -, 120, 7,175

Oymphia Sullivan, 160, 240, 1200, 90, 600

Elijah Padgett, 50, 30, 240, 15, 300

Joy (Ivy), McDaniel, 15, 65, 175, 12, 150

Bryant McDaniel, 30, 50, 50, 10, 100

John McDaniel, 15, 65, 320, 17, 90
James Gay, 25, 15, 150, 6, 110
Richard B. Carlton, 21, 19,170, 12, 150
John Stewart, 35, 75, 400, 20, 300
John E. Pledger, 17, -, 170, 10, 100
Nath. Dykes, 23, -, 230, 16, 126
Kindred Jones, 36, -, 160, 12, 100
Hugh Spears, 300, 1200, 6000, 60, 1000
Frederick Swearingen, 29, -, 145, 10, 260
John P. Lockey, 80, 160, 1000, 90, 330
John G. Russ, 350, 570, 6000, 175, 1500
Robert Gilbert, 35, 45, 400, 20, 275
Joseph T. Russ, 300, 440, 6000, 140, 1600
Aaron Underwood, 30, -, 175, 9, 190
Andrew Williams, 32, 48, 275, 20, 190
David Williams, 20, -, 100, 6, 90
Jery Davis, 15, 55, 125, 14, 80
Luke Sumerlin, 10, 70, 120, 7, 116
Jerry P. Personale, 30, -, 150, 9, 320
Elizabeth Garner, 12, 72, 120, 7, 98
Irwin Miller, 26, -, 110, 9, 85
Enoch Miller, 20, -, 120, 8, 75
William Jackson, 20, -, 100, 15, 185
Amos Snell, 100, 800, 4500, 175, 1300
Green Snell, 40, -, 200, 20, 250
Wm. N. H. Harrell, 20, 60, 140, 15, 210
Dangold Anderson, 20, 80, 300, 18, 160
Robt. T. Smith, 25, 110, 460, 10, 85
Abram Duncan, 30, 50, 150, 25, 125
Isaiah Williams, 25, -, 100, 14, 100
Watson Davis, 40, 40, 240, 27, 190
Harrison Stinson, 90, 30, 360, 45, 395
John B. Brown, 240, 760, 5000, 190, 1800
Lucretia Britt, 80, 80, 800, 70, 275

Benjamin Hays, 25, 375, 4000, 20, 700
Charles Slater, 30, 50, 300, 62, 150
Joseph W. Russ, 500, 300, 4000, 190, 1400
Saml. Stephens, 60,100, 350, 65, 360
James D. Stephens, 30, - 173, 45, 200
Riley Dykes, 25, -, 100, 7, 100
Simmons J. Baker Sr., 800, 1500, 12500, 375, 2000
F. R. Ely, 650, 900, 9156, 550, 2700
John P. Mayo, 14, -, 120, 47, 450
James T. Mayo, 14, -, 120, 47, 420
James Williams, 150, 170, 1300, 120, 440
Wylie Stewart, 30, -, 100, 19,-
Wm. F. Snelling, 160, 300, 1600, 125, 800
J. H. Britt, 400, -, 2500, 250, 1000
Thomas Clayton, 40, -, 310, 64, 320
John Redd, 20, -, 100, 16, 300
Wm. R. Petteray (Pettesay), 60, 20, 320, 90 295
Richard Barnes, 35, 45, 800, 60, 310
Jacob J. Pelt, 70, 40, 600, 16, 260
Wynn's Estate, 500, 200, 5000, 190, 1400
C. Whitaker, 300, 660, 4000, 220, 1300
George Gray, 500, 300, 6000, 320, 1600
John Waddill, 350, 1250 14000, 190, 4000
Archibald Patterson, 40, 40, 240, 60, 375
Patrey Gilbert, 40, 160, 600, 55, 800
Washington Geloat, 60, 160, 600, 90, 450
Benjamin Holder, 100, 160, 1100, 110, 460
Isaac Widgeon, 350, 650, 5000, 275, 2500
Richard Wilson, 30, 50, 350, 45, 126
James L. G. Baker, 470, 300, 8000, 250, 1900

Wm. Nickels, 240, 300, 7000, 400, 1200

John Gay, 40, -, 175, 38, 125

Armistead S. Rains, 50, 30, 400, 60, 460

Jesse Robinson, 400, -, 2800, 175, 2000

Benjamin Wynns, 800, 1000, 10000, 300, 2500

William Vickery, 30, -, 100, 15, 150

Jane Vickery, 20, -, 100, 17, 250

M. C. Ross, 20, -, 100, 28, 120

Mary D. Watson, 60, 260, 1500, 110, 275

Thomas Wilton, 40, 160, 1200, 30,175

Edum Whitehead, 180, 150, 2000, 400, 900

John Tanner, 200, 730, 6000, 380, 900

James Kent, 40, 40, 200, 40, 900

William Barksdale, 60, -, 500, 40, 140

John Boggs, 22, 58, 240, 60, 200

John Davis, 120, 80, 1700, 65, 950

Edward C. Belamy, 2000, 2000, 30000, 2000, 5000

Benjamin Myles (Myler), 15, -, 150, 12, 160

James Hall, 90, 70, 400, 35, 280

John S. Mears, 18, 32, 160, 7, 65

Levin Gillstrap, 30, -, 150, 19 200

Arhibald Ball, 30, -, 175, 14, 300

Asa Shiver, 25, 55, 240, 11, 140

Charles Howard, 65, -, 250, 40, 365

John M. Caraway, 40, -, 400, 45, 95

Elias Hammonds, 45, -, 375, 16, 130

Samuel Jones, 75, 5, 600, 90, 300

Simmon J. Baker Jr., 500, 700, 8000, 300, 1600

Bayliss Purson, 35, 45, 240, 60, 320

F. B. Calloway, 40, 40, 320, 50, 30

Isaiah Daniels, 100,780, 3000, 110, 700

Elizabeth Calloway, 30, 10, 200, 60, 195

William N. Matthews, 110, 10, 800, 65, 375

James Warmouth, 30, 80, 500, 15, 200

John Smith, 100, -, 400, 90, 100

John Wadford, 40, 80, 800, 70, 130

Sarah Marton, 45, 80, 800, 70, 130

Elizabeth Brantley, 100, 300, 1200, 75, 340

Tabitha Britt, 30, 730, 5000, 16, 95

Stephen Lundy, 175, -, 1000, 96, 450

Nicholas W. Miller, 16, -, 100, 40, 140

Wm. A. Whitefield, -, -, -, -, 100

Benjamin Whitefield, 75, -, 400, 16, 75

William Fountain, 23, -, 125, 12, 100

William Matox, 80, -, 100, 70, 120

John Q. Lundy, 20, -, 120, 16, 90

Wm. A. Lundy, 30, -, 110, 12, 75

Robert A. Young, 400, 260, 4500, 190, 1250

William Hall, 260, 160, 3000, 140, 1350

Jonathan W. F. Jenkins, 70, -, 400, 120, 200

W. A. Abercrombie, 110, -, 320, 100, 400

Bryant Pittman, 14, 316, 1000, 21, 116

Moses Jenkins, 30, 50, 100, 46, 230

Laban Beauchamp, 20, 60, 120, 16, 85

Noel Grantham, 60, 20, 240, 24, 100

Henry A. Bright, 140, 280, 2000, 110, 500

S. W. Smith, 375, 273, 3000, 275, 1400

Thomas W. Edwards, 125, 120, 725, 20, 478

Philemon Conelly, 50, 30,125, 15, 190

John Conelly, 40, 40, 130, 9, 185

John Callutter, 25, 65, 125, 8, 90

John Melton, 320, 2434, 3200, 375, 2000

John W. Pouse (Poner), 60, -, 300, 25, 200

Elias Wester, -, -, -, -, -

William Carpenter, 30, 50, 500, 12, 173

George C. Neely, 15, -, 150, 7, 2000

William Williams, 20, -, 100, 6, 75

James McG. Hunter, -, -, -, -, -

Wylie J. Faircloth, 20, 180, 100, 9, 450

Noah Faircloth, 30, -, 300, 12, 195

E. M. Skipper, 45, 35, 1200, 15, 220

Henry Sellers, 20, -, 200, 10, 80

Jerry Sims, 200, 700, 2700, 115, 1200

Samuel Gammon, 85, 120, 8500, 25, 450

Blackledge Bryan, 50, -, 50, 20, 135

Jno. W. & W. P. Gammon, 32, -, 320, 10, 273

Edward Bryan Sr., 180, 300, 1800, 175, 950

Thomas Barnes, 510, 850, 5100, 400, 1600

Thomas Godwin, 180, -, 1800, 175, 750

Edwd. Bryan Jr., 160, -, 1600, 25, 840

William Sims, 85, 115, 180, 65, 475

F. R. Pittman, 200, 275, 1500, 100, 600

Mavlina F. Lott, 120, 520, 3000, 40, 450

Boling B. Barkley, 100, 220, 1500, 130, 375

Waller(Walter) J. Robinson, 280, 200, 3000, 175, 1165

Thos. J. Anderson, 50, 150, 150, 20, 125

William McNeely, 250, 570, 6000, 300, 2500

Saml. B. Williford, 40, -, 100, 20, 260

James O. Banter (Baxter), 100, 580, 4000, 30, 195

Solomon Banter, 25, -, 100, 9, 80

William Ferguson, 20, -, 150, 12, 60

Israel A. Banter (Baxter), 30, -, 150, 10, 180

Robert Kidd, 25, -, 125, 8, 164

Edward Minchen, 80, -, 800, 20, 461

John McDaniel, 60, -, 600, 60, 510

Richard Bell, 35, 45, 200, 50, 375

David Baxter, 65, -, 200, 20, 700

Robert McDaniel, 30, 10, 200, 17, 275

Henry McDaniel, 20, -, 100, 10, 110

William Trewathan (Trevathan), 60, 300, 400, 20, 140

Johnson Alsobrook, 65, 55, 600, 37, 270

M. B. Pender, 100, 275, 1200, 90, 475

Jno. A. Sypett (Syfrett), 60, 220, 1800, 40, 380

Elijah Bryan, 400, 2100, 18000, 300, 2600

William E. Harvey, 160, 200, 2000, 125, 568

John Durham, 35, 165, 600, 15, 450

James Dykes, 100, 180, 1500, 95, 700

James Bomcastle, 60, -, 300, 25, 120

Mary Roberts, 250, 2230, 12400, 375, 2200

Wm. B. Dickson, 200, 40, 1000, 250, 700

Eleazer Watts, 50, 30, 500, 27, 100

N. Holder, 45, -, 225, 15, 195

L. H. Hearn, 140, 20, 480, 300, 900

Wm. Rawls, 30, 30, 300, 20, 90

Emily Dickson, 270, 1500, 3500, 275, 1000

John S. Harvey, 65, -, 650, 70, 320

John Bryan, 50, 30, 700, 25, 315

David Blackshear, 225, 175, 1600, 195, 1300

Marmaduke N. Dickson, 280, 440, 5000, 420, 1250

Alexander M. Raine, 40, -, 400, 5, 375

N. A. Long, 720, 1400, 16000, 390, 3000

Aquilla Bonn, 30, -, 30, 27, 100

Joseph B. Evrett, 100, 20, 400, 40, 570

John Padgett, 40, -, 200, 26, 500

Ethington J. Merrit, 180, 100, 1000, 175, 800

Isaac Wimberly, 30, 90, 650, 24, 390

William Sarey, 60, 340, 1200, 90, 460

Thos. M. Bush, 1000, 400, 5500, 790, 4000

Joseph W. Russ, 400, 400, 3000, 475, 1700

Richard Rankin, 20, 149, 470, 96, 340

Benjamin Darden, 90, 40, 700, 40, 300

John G. Routher Sr., 850, 900, 15000, 560, 5480

Rebecca Hill, 80, 120, 1000, 65, 200

John Butt Sr., 200, 440, 300, 90, 800

Jonathan Jones, 15, 65, 150, 15, 190

Ephraim C. Haynes, 30, -, 120, 10, 200

D. W. Horne, 300, 660, 6000, 275, 2500

Jacob T. Porter, 50, 230, 1200, 60, 195

Richard McDaniel, 17, -, 170, 10, 250

Thos. Cook, 50, -, 300, 20, 190

Ferdinand Weeks, 20, -, 100, 7, 210

Alfred Nickels, 25, -, 100, 6, 120

James Scott, 10, 70, 300, 12, 80

Harrison Hare, 30, -, 100, 10, 120

F. P. Haywood, 250, 700, 1000, 100, 1000

Alfred Basil, 20, 60, 100, 7, 75

Isham H. Basil, 35, 45, 150, 13, 370

B. F. Wood, 140, 160, 1200, 30, 2000

Wm. M. Owens, 150, 850, 1500, 60, 6225

Eveline Jordan, 70, -, 700, 15, 300

John Hart, 19, -, 190, 12, 190

Alfred Collins, -, -, -, -, -

Elizabeth Brown, 150, 150, 1500, 40, 1000

Martin Nobles, 15, 21, 150, 8, 90

William Boon, 40, -, 400, 15, 240

Alexander Johnson, 15, -, 150, 7, 50

Wm. C. Neil, 250, 150, 1500, 20, 1000

Benjamin Neil, 20, -, 200, 11, 175

James A. Neil, 20, 100, 120, 10, 150

Henry Bell, 25, 75, 100, 7, 120

Whitmell Boon, 20, 60, 115, 10, 95

Samuel Oswald, 50, -, 200, 20, 225

Green B. Yarborough, 45, -, 100, 9, 193

Perry J. Lockhart, 25, -, 120, 10, 65

Nancy Allen, 20, 60,100, 7, 90

Wilson Carlisle, 26, 95, 130, 16, 2770

Freeman Irwin, 50, 30, 100, 14, 400

John P. Martin, 50, 30, 100, 17, 30

William Coonrod, 65, -, 175, 20, 380

Allen Irwin, 10, -, 100, 10, 148

N. R. McAnulty, 35, 55, 125, 16, 125

Peter Taylor, 43, 35, 100, 12, 300

Hardy Hart, 40, -, 300, 9, 764

Jesse Hart, 40, -, 175, 6, 75

Israel Baxter (Banter), 55, 105, 200, 14, 1200

Thomas S. Westin, 25, -, 180, 7, 127

John W. Staley, 14, 100, 1?0, 4, 90

Edwards & Dickinson, 180, 260, 1800, 200, 1000

Geo. W. Tillinghart, 150, 400, 1500, 200, 1200

Jefferson County, Florida
1850 Agricultural Census

The University of North Carolina at Chapel Hill filmed the 1850 agricultural census for Jefferson County from originals at the Florida State University under a grant from the National Science Foundation in 1963.

Columns 1, 2, 3, 4, 5, and 13 represent the following information on the census:
1. Name of Owner, Agent or Manager of Farm
2. Acres of Improved Land
3. Acres of Unimproved Land
4. Cash Value of the Farm
5. Value of Farming Implements and Machinery
13. Value of Livestock

David B. Hart, 100, 220, 800, 30, 495

James Clack (Clark), 14, 26, 80, 15, 50

Levi Massey, 60, 60, 300, 50, 295

Rollin Reaves, 50, 110, 350, 100, 475

Wiley Barwick, 50, 30, 250, 20, 290

Mary Folson, 50, 30, 250, 20, 290

Major Surles, 80, 740, 400, 30, 635

Francis Shepherd, 40, 30, 200, 10, 175

Thomas Shepherd, 23, 57, 125, 20, 160

Kindred Sauls (Souls), 100, 357, 500, 25, 250

Daniel Howell, 20, -, -, 9, 145

Jessee Cone, 50, 35, 300, 30, 150

Lewis Cone, 80, 160, 500, 60, 392

Aylesbery B. Lord, 80, 80, 500, 60, 325

Vincent Tanner, 25, -, -, 10, 140

Joseph Tanner, 80, 80, 200, 35, 40

Nicholas Fennel, 125, 295, 900, 230, 585

Joshua Presseley, 40,-, -, 220, 500

Benjamin Umphreys, 43, 117, 215, 50, 210

John Williams, 55, 25, 400, 25, 165

Hardy Whittey, 30, -, -, 10, 105

Henry Saunders, 60, 180, 600, 135, 540

John Mikell, 20, 60, 200, 45, 470

William Mikell, 24, 56, 200, 10, 290

Elias E. Blackburn, 120, 20, 300, 85, 396

John J. Johnson, 25, 95, 150, 50, 130

Charles Barronton, 66, 414, 300, 20, 220

James A. Cooksey, 125, 115, 500, 90, 410

Daniel Pond, 30, 50, 180, 40, 210

John McEaker, 20, -, -, -, 150

John Hurst, 80, 80, 400, 10, 200

Catherine Decausey, 44, 40, 200, 5, 265

Charles Fox, 80, 80, 500, 10, 410

George W. Blackburn, 90, 70, 400, 15, 297

Sterling Sims, 120, 240, 500, 445, 700

Thomas V. Clark, 55, 25, 150, 45, 230

Joshia Taylor Jr., 35, -, -, 10, 218

John G. Powell, 30, -, -, 35, 116

George W. Taylor, 80, 104, 900, 200, 800

Norman Williams, 133, -, -, 10, 167

James S. Russell, 100, 324, 600, 100, 494

Lewis Williams, 30, 130, 150, 10, 120

John M. Holder, 90, 70, 200, 10, 305

Green Faglio, 45, 35, 150, 10, 236

Malinda Colson, 130, 350, 700, 210, 898

John J. Edwards, 20, 140, 200, 35, 260

Richard Taylor, 80, 140, 1500, 110, 420

Elisha Bozeman, 25, 95, 150, 30, 140

William G. Clark, 200, 430, 2000, 600, 1700

Jams Slater, 300, 390, 2000, 480, 1150

William J. Woods, 160, 440, 600, 200, 640

James H. Taylor, 100, 140, 300, 125, 673

Ann Chesnutt, 12, -, -, 10, 120

Baptist Bramby, 100, 76, 600, 60, 1480

Benajah Edwards, 20, -, -, 10, 114

Robert C. Hurst, 170, 270, 1200, 500, 1510

Firgy Howell, 35, -, -, 50, 385

Jacob Rickert, 70, 90, 500, 100, 800

Arnold Cone, 100, 60, 400, 45, 325

Francis Hay, 60, 140, 300, 35, 245

William Snead, 200, 120, 1500, 175, 780

William Mershon, 40, 120, 125, 100, 300

Kidder M. Moore, 300, 500, 2060, 600, 1550

William V. Ramsey, 20, -, -, 15, 170

Christopher Colman, 18, -, -, 10, 120

John A. Hallman, 400, 650, 3000, 600, 1410

Joel Moore, 30, 210, 400, 50, 328

Robert Jones Sr., 80, 40, 500, 40, 357

Ann B. Sloan, 80, 120, 800, 180, 775

Robert H. Gamble, 1541, 6590, 14000, 4000, 4500

Elizabeth Wirt, 1000, 2080, 6000, 700, 2570

Smith Simkins, 250, 710, 3000, 400, 910

Hartwell Watkins, 40, 200, 200, 50, 186

Allen Windham, 65, 275, 300, 50, 340

Mary Saltenstall, 35, 721, 400, 50, 293

Isaac Rouse, 130, 125, 800, 75, 800

William D. Moseley, 850, 1655, 8500, 1000, 2652

John N. Partridge, 100, 300, 1500, 150, 300

Samuel Davis, 26, 54, 150, 25, 225

Jonathan Bentley, 25, 65, 160, 10, 280

William Dawkins, 40, 160, 200, 10, 240

James R. Brooks, 70, 170, 1200, 90, 555

Benito Aguero, 10, 32, 600, 25, 70

William Gorman, 350, 450, 8000, 500, 1500

Charles D. McClellan, 50, 70, 1000, 75, 300

Burner Hay, 18, 37, 70, 5, 90

Margaret E. Stephens, 20, 60, 400, 30, 576

John P. Cooksey, 100, -, 400, 100, 380

Samuel W. Neely, 20, -, -, 10, 170

John Stakeley, 12, -, -, 15, 70

B. H. Ward, 20, 60, 100, -, 140

A. R. McCall, 35, 205, 350, 10, 500

Richard Long, 275, 1385, 2500, 580, 1170

Sarah Grantham, 90, 210, 450, 30, 270

Robert Potts, 27, 23, 270, 5, 150

Rebecca Grantham, 110, 180, 770, 30, 960

Robert Long (Lang), 40, 120, 400, 80, 670

Oliver Ham (agt TR), 750, 4030, 6750, 850, 1900

Charles Martin, 20, -, 80, 20, 160

Vincent, J. Strickland, 160, 200, 2400, 1000, 2135

Gideon Mills, 30, 50, 300, 20, 356

Mary Richardson, 20, 20, 200, 5, 200

William Jones, 35, 5, 500, 50, 95

Samuel C. Croft, 10, 40, 100, 5, 100

Valentine Rowell, 25, 15, 125, 44, 180

John L. Taylor, 700, 440, 2000, 1400, 1180

Isaac Townsend, 80, 240, 800, 200, 1700

Calvin Davis, 110, -, -, 100, 530

John G. Plant (agt JGA), 700, 800, 8400, 1750, 3000

Benjamin Farrel, 23, -, -, 10, 220

James Gaskins, 16, 24, 164, 20, 340

Cornelius Beasley, 200, 540, 2500, 1220, 1225

Josiah P. Scruggs, 70, 250, 710, 137, 700

John G. Bevill, 120, 120, 720, 90, 470

Washington Floyd, 800, 2640, 6400, 945, 2450

Joshia Floyd, 30, 90, 300, 50, 200

Hannah High, 23, 137, 115, 35, 310

Paul H. Harley, 100, 540, 500, 175, 490

Asa May, 60, 260, 1200, 740, 670

Caroline G. Cole, 350, 2070, 5450, 80, 2030

William H. Arendal, 270, 10, 2500, 60, 915

Joseph C. Reid, 30, 50, 300, 100, 250

William H. Stafford, 28, -, -, 150, 600

James Kersey, 40, 40, 280, 50, 270

James S. Bond, 200, -, -, 200, 890

Gage W. Gelzer, 150, 610, 1500, 700, 1120

Joseph McCant, 130, 470, 925, 60, 340

Est. of T. P. Randolph, 480, 80, 1440, 230, 520

Thomas Powell, 100, 100, 200, 50, 690

John M. Ragson, 170, 470, 340, -, 108

Jane S. Walker, 150, 210, 2000, 415, 770

Hellary Whitehurst, 50, 190, 400, 50, 525

Absolem Whitehurst, 25, 15, 90, 5, 238

Lewis Ivy, 30, 50, 150, 30, 165

Mitchel L. Scott, 60, 190, 1000, 260, 910

John Slaughter, 40, 160, 400, 30, 890

Alexander Scott, 112, 288, 896, 450, 1180

Est. of A. B. Scott, 65, 100, 240, 10, 150

James Johnson, 40, -, 300, 20, 46

John Huggins, 10, -, 50, 10, 105

Martin Bishop, 17, 63, 85, 20, 170

Pearson D. Bridges, 20, -, 200, 15, 145

David Bishop, 55, 105, 550, 20, 525

James L. Kirkland, 17, -, 100, 5, 118

Henry Walker, 65,175, 455, 80, 930

Isaac L. Lamb, 40, -, -, 50, 200

George Bishop, 36, -, 80, 5, 245

Jessee Bishop, 60, 60, 30, 80, 400

David Osborn, 12, 38, 60, 30, 125

Joshia McCann, 28, -, -, 10, 140

Henry Goodman, 30, 130, 120, 30, 150

William Powell, 50, 110, 150, 50, 255

LittleBery Walker, 12, 38, 80, 5, 170

Joseph W. Clayton, 35, -, -, 50, 370

John Arnold, 18, -, -, 20, 225

John Holton, 18, -, -, 50, 130

Joel Walker, 50, 130, 700, 170, 370

David Walker, 50, 130, 300, 10, 130

Stephen Lightsey, 100, 100, 900, 130, 690

Andrew J. Whitehurst, 40, 80, 300, 100, 465

Hampden S. Linton, 350, 490, 3000, 500, 1200

Martin Sparks, 25, 55, 150, 10, 75

Whitmell Horton, 22, 18, 150, 35, 338

Martha Henry, 35, 40, 120, 130, 373

James Walker, 80, 320, 500, 100, 540

Joseph Kinsey, 80, 40, 400, 150, 760

Jehu Grubbs, 20, -, -, 5, 130

Jesse Kinsey (Kirsey), 31, 49, 219 40, 432

John Kirsey, 25, 15, 200, 10, 260

Nancy Murphey, 20, 20, 100, 5, 195

Josiah Knight, 80, 80, 4000, 100, 370

Benjamin Grubbs, 65, 15, 325, 75, 510

Samuel Grubbs, 23, -, -, 5, 45

William H. Scruggs, 300, 600, 1800, 400, 1680

John Manning, 35, -, -, 35, 266

Richard H. Shackleford, 35, 45, 200, 30, 200

Daniel Slaughter, 25, -, -, 40, 145

John Foglie (Faglie, Faglio), 70, 250, 500, 130, 800

James Ramsey, 100, 220, 1000, 55, 180

Tempy Woolf, 70, 90, 350, 5, 200

Agnes Woolf, 30, -, 120, 10, 240

Ann Woolf, 30, -, 120, 5, 140

Nelson Rayfield, 75, -, 300, 75, 560

George W. Standley, 45, 153, 225, 80, 270

William Standley, 35, 125, 350, 100, 240

Mary Johnson, 200, 220, 1500, 590, 1445

Cudger Bellamy, 40, -, 200, 35, 310

James Williams, 100, 220, 800, 60, 540

Josiah Scoggins, 50, 100, 250, 15, 260

Wilkins C. Smith, 160, 420, 1000, 400, 1100

Elizabeth Jenkins, 36, 84, 144, 24, 220

Granderson Barker, 30, -, -, 5, 120

Unity Alman, 16, 24, 120, 35, 160

John Gamble, 50, -, -, 100, 505

John T. Smith, 34, 45, 100, 30, 140

Levi Leopard, 30, 50, 65, 10, 100

John Bellamy, 100, 220, 800, 415, 890

William Kersey, 75, 305, 800, 410, 630

Isaiah Hampton, 25, -, 100, 25, 275

Robert D. Johnson, 275, 445, 1300, 500, 1000

Newton Mathers, 150, 262, 2050, 200, 665

Seaborn Jones, 30, 10, 150, 100, 820

Robert Jones Jr., 40, 80, 200, 15, 250

William Daricott, 35, 85, 250, -, 270

Thomas Stewart, 45, 145, 180, 115, 250

Joshua Clark, 40, -, 200, 80, 240

Luke Bozeman, 30, 90, 150, 25, 285

John Moore, 100, 220, 1000, 160, 555

John G. Peak, 37, 93, 200, 100, 1070

Banister Kersey, 80, 237, 800, 80, 670

John White, 33, 58, 350, 5, 210

Asa Anderson, 23, 137, 300, 10, 200

Daniel T. Lingo, 100, 20, 1000, 110, 730

John K. Vann, 30, 130, 200, 5, 160

Willis T. Davis, 60, 10, 400, 100, 210

John Wooten, 100, 260, 800, 25, 515

Hardy Moore, 100, 220, 1000, 500, 1060

William Albritton, 60, -, -, 25, 585

Franklin Faglie, 65, 956, 300, 50, 290

Samuel Stokes, 80, 80, 500, 300, 1545

William Horton, 65, 15, 200, 20, 270

John M. McMillan, 100, 880, 500, 40, 500

Mitchel B. Umphreys, 60, 100, 350, 50, 550

Richard Gilbert, 80, 160, 500, 35, 210

William Chesnutt, 60, 20, 300, 60, 300

Barnhard Loeb, 100, 300, 1000, 170, 1060

John Anderson. 35, -, -, 10, 225

David Anderson, 100, 73, 700, 50, 580

Feady Litchworth, 33, 47, 150, 30, 285

Elias Lastinger, 80, 240, 500, 45, 1120

Green Sledge, 50, 140, 400, 75, 375

Malachia Goodman, 20, -, -, 30, 275

William Watson, 25, 55, 150, 45, 350

John Braden, 200, 300, 1500, 680, 648

John Scott, 40, -, -, 10, 120

John O. Herring, 18, 62, 150, 25, 105

Daniel H. Shepard, 16, 64, 80, 25, 100

Ally Fife, 20, 20, 100, 10, 130

Andrew J. Moore, 40, 280, 200, 35, 260

Ashford Jenkins, 50, 110, 350, 35, 348

Joseph Joyner, 70, 10, 350, 45, 520

Ann Hart, 45, 75, 225, 15, 220

Robert Bryan, 25, -, -, 35, 180

Needham Bryan, 150, 410, 1200, 470, 830

Daniel Bryan, 30, -, -, 50, 305

Alvin May, 100, 620, 4000, -, 470

Fetnah Cooksey, 250, 230, 2000, 4500, 1000

Martha P. Triplett, 160, 240, 2400, 5000, 1210

Dorothy Triplett, 160, -, 960, -, 320

Thomas Triplett, 160, -, 960, 25, 320

Paul Ulmer, 350, 520, 6100, 500, 1285

James Kirksey, 600, 300, 3600, 500, 1130

Lewis Lynn, -, -, -, -, 200

John Kirksey, -, -, -, -, 320

William R. Taylor, 200, 360, 5000, 300, 1000

Ada Winck, 375, 1385, 9000, 100, 800

Sophia Ellis, -, -, -, -, 7

John L. Houck, -, -, -, -, -

William Hollingsworth, 475, 965, 7200, 800, 1600

Daniel D. Sturges, 200, 500, 5000, 500, 900

Samuel H. Whitaker, 600, 180, 1000, 20, 500

John Grantham, 50, 350, 2000, 20, 500

Edward Willey, 375, 375, 3000, 200, 1100

James B. Edward, 40, 40, 400, 180, 1000

Elizabeth Woods, 40, 40, 300, 20, 300

Henry R. Edwards, 270, 650, 5500, 450, 900

William S. Hill, 200, -, -, 200, 700

Wesley Adams, 150, 250, 1600, 250, 600

George Jones, 1400, 2600, 21600, 1000, 4200

Jesse Hightower, 70, -, 350, 10, 350

Mary Edwards, 12, 28, 100, 5, 150

Ann Christopher, 150, 130, 840, 150, 450

Hezekiah Tucker, 80, -, 320, 30, 250

John Buckle, 80, 80 800, 70, 300

Irwin Grainger, 25, 55, 300, 10, 300

Timothy T. Smith, 80, -, 200, 20, 200

James A. Mills, 15, 65, 150, 30, 150

Charles Adams, 10, -, 50, -, 130

Henry Mills Sr., 160, 40, 2800, 50, 470

Stephen Vickers, 20, -, 100, 5, 130

E. _. Proctor, 17, 146, 160, 5, 200

Robert Clark, 12, 78, 200, 5, 100

Samuel Gale, 30, 160, 400, 100, 270

Charles G. Bell, 10, 70, 590, 75, 200

Guilford W. Houston, 40, 40, 600, 15, 140

William Thomas, 20, 160, 1200, 50, 375

John A. Edwards, 130, 700, 17800, 570, 1100

Wilson Barronton, 35, 125, 1000, 40, 400

John B. Page, 200, 360, 5600, 450, 1250

Everit Sealy, 160, 200, 1800, 100, 850

John Grainger, 40, 200, 1200, 40, 520

Maria Ulmer, -, 35, 264, -, 70

Everit Lee, 18, 3, 210, 5, 112

James Casaday, 10, 45, 275, 40, 240

Julius High, 70, 170, 900, 160, 430

Pinckney Bellinger, 18, 12, 200, 10, 400

Octavius H. Gadsden, 500, 500, 5000, 1000, 1700

Octavius H. Gadsden (agt), 800, 1700, 12500, 1500, 2600

Octavius H. Gadsden (agt), 700, 1740, 10700, 1200, 1940

H. W. White agt of A. M. 500, 1900, 12200, 550, 1620

William Wooten, 30, 210, 1200, 50, 350

John Hargrove, 40, 40, 600, 100, 550

William S. Murphey, 300, 670, 6500, 500, 1000

Lewis Cannon, -, -, -, -, 150

Mannassah Micheau, 140, 340, 2200, 300, 684

William T. Harrison, -, -, -, -, 150

Charles W. DeWitt, 50, 65, 690, 75, 430

John DeWitt, 50, 65, 690, 25, 350

William H. DeWitt, 50, -, 125, 30, 310

Stephen D. Strickland, 15, -, 50, 25, 310

Sarah Ellis, 100, 100, 1000, 50, 675

David P. F. Newsom, 10, 650, 6000, 150, 560

James Brumby, 12, -, 50, 5, 80

John McLean, -, -, -, 100, 200

William G. Powell, 40, 40, 200, 50, 230

Elias Sealy, 15, -, 45, 10, 130

George S. Armear, 12, -, 40, 10, 300

John Freeman, 30, 150, 3600, 50, 300

John Seaver, 14, -, 20, -, 50

James Brooks, 30, -, -, 100, 95

Abner Wethington, 125, 275, 4000, 450, 625

John Q. Wethington, 140, 340, 2600, 40, 360

Alexander Moffatt, 25, 55, 320, 20, 150

Council Wooten Sr., 160, 640, 7000, 360, 1100

Sarah Grainger, 35, 85, 840, 10, 120

Henry Grantham, 50, 30, 240, 5, 170

Caroline S. Johnston, 15, 5, 200, 5, 140

Lewis L. Gregory, 45, 115, 1000, 35, 270

John M. Hooks, 60, 180, 700, 30, 340

Abraham A. Wilford, 80, 280, 800, 50, 290

Francis P. Miller, 120, 360, 1000, 410, 1180

Sarah Clark, 90, 70, 600, 75, 930

Noah H. Teat, 230, 170, 2500, 1000, 1850

Samuel Bently, 50, -, -, 45, 250

Nathaniel Turner, 150, 90, 1500, 200, 760

Job Wilford, 100, 220, 800, 100, 730

Edmund Davis, 150, 430, 1000, 280, 830

Charles C. T. Singleton, 600, 1080, 4000, 725, 1360

John Doggett, 550, 650, 3500, 1000, 2950

Joshua Taylor Sr., 120, 380, 700, 530, 645

Benjamin Rogers, 38, -, -, 45, 290

John Fuquar, 25, -, -, 10, 200

John Dawkins, 60, 20, 300, 130, 416

Francis Dawkins, 100, 340, 1000, 125, 635

James Scott, 400, 500, 3000, 500, 1620

Uriah Brown, 85, -, -, 50, 525

Martha Rogers, 20, -, -, 30, 150

William Bailey, 1600, 2000, 8000, 2775, 10020

John Johnson, 240, 190, 1920, 500, 1240

Mary McN. Morris, 110, 210, 900, 200, 1000

Benjamin Windham, 110, 210, 1100, 200, 810

Eligah Richardson, 25, 135, 150, 70, 225

John Lee, 40, 120, 300, 40, 90

Farmer Lee, 20, 140, 150, 10, 107

Charles Jolly, 20, -, -, 50, 95

Britton Edwards, 65, 95, 350, 40, 315

William J. Bailey, 1250, 3510, 12000, 1320, 4450

John Finlayson, 1000, 300, 10000, 1400, 3750

Richard Shackleford, 40, 160, 400, 50, 250

Henry Demott, 15, 25, 150, 25, 200

Harvy G. McClellan, -, -, -, -, 160

Crumala A. Bellamy, 400, 600, 4000, 900, 1600

Thomas Townsend, 1000, 7000, 10000, 1500, 3960

Joseph Dawkins, 1000, 800, 2000, 1500, 2800

William H. Ware, 348, 735, 8000, 1000, 2350

Grace Ware, 100, 125, 2000, 140, 350

James R. Tucker, 500, 920, 7500, 700, 3000

Daniel Bird, 750, 1108, 11250, 500, 2156

Zachariah Bailey, 200, 200, 3000, 1000, 2100

Ellen A. Beatty, 1000, 2500, 4000, 1000, 2700

Benjamin Johnson, 300, 420, 3000, 600, 1570

John G. Methus, 285, 315, 2800, 600, 1200

John G. Pettus, 200, 960, 2500, 300, 100

Martin Palmer, 700, 1360, 10000, 2000, 3285

William West, 125, 115, 500, 300, 900

James T. Baker, 60, -, -, 30, 265

Hugh Cornel, 30, -, -, 30, 260

Frances Cuthbert, 40, 40, 150, 40, 140

William H. Andrews, 20, 20, 150, 50, 300

Charles L. Powell, 200, 400, 2500, 300, 1250

Leon County, Florida
1850 Agricultural Census

The University of North Carolina at Chapel Hill filmed the 1850 agricultural census for Leon County from originals at the Florida State University under a grant from the National Science Foundation in 1963.

Columns 1, 2, 3, 4, 5, and 13 represent the following information on the census:
1. Name of Owner, Agent or Manager of Farm
2. Acres of Improved Land
3. Acres of Unimproved Land
4. Cash Value of the Farm
5. Value of Farming Implements and Machinery
13. Value of Livestock

A. O. Hock, 25, 40, 2000, 50, 125
John P. Duval, 80, 500, 580, 50, 930
James G. Sweat, 40, 8, 100, 10, 100
L. A. Thompson, 180, 300, 1000, 25, 1000
Philip T. Pearce, 20, -, 250, 10, 300
James W. Dabney, 130, 70, 2500, 120, 1003
David L. Evans, 40, 120, 800, 10, 475
Fred Farrior, 25, 95, 500, 8, 100
R. B. Carpenter, 60, 100, 2000, 430, 530
Dan B. Fisher, 100, 50, 700, 50, 400
John B. DeCorce, 25, -, 1500, 10, 250
Daniel G. Horn, 31, 10, 200, 10, 50
A. A. Fisher, 115, 245, 5400, 1000, 3925
Hiram Wheeler, 33, 35, 350, 10, 140
James H. Branch, 350, 370, 7200, 300, 2080
M. M. Felkell, 10, 9, 850, 10, 234
Dan Houck, 125, 208, 2664, 315, 545
James Cohron, 25, 15, 300, 30, 100
Charles B. Weste, 80, 95, 1000, 100, 600
N. W. Cason, 100, 600, 10000, 500, 1500

A. M. R. Sessions, 130, 320, 2900, 300, 1000
W. P. Scott, 30, 210, 1000, 45, 300
James Courtney, 35, -, 150, 5, 160
G. W. Philips, 50, 80, 500, 50, 250
M. Strickland, 100, 60, 1000, 25, 260
A. K. Butler, 270, 95, 2920, 200, 710
R. Van Brants Jr., 275, 452, 5600, 500, 810
D. Williams, 120, -, 500, 20, 444
James R. Nix, 155 350, 3000, 300, 900
John Harley, 150, 330, 2000, 100, 430
M. Hukins, 135, 25, 1000, 50, 440
Jos. W. Bennerman, 600, 200, 4000, 300, 1700
W. A. Care, 600, 1400, 20000, 325, 2000
W. Roberts, 140, -, 1400, 250, 2405
F. R. Cotton, 700, 1280, 20000, 1000, 2530
Elijah Perkins, 40, 120, 1000, 25, 500
William Rowland, 40, 40, 520, 20, 165
David Young, 150, 145, 1500, 240, 780
Charles Jones, 35, 60,700, 15, 320
Nicholas Lloyd, 60, 60, 600, 50, 538
Laban Rawls, 35, 150, 250, 30, 405

Joshua Spears, 60, 100, 500, 30, 100
Benjamin W. Guest, 25, 15, 100, -, 145
John F. Williams, 18, 10, 200, 30, 130
Lovick Green, 15, 10, 100, 15, 660
William Ash, 22, 60, 300, 25, 200
Seaborn Rawls, 150, 150, 700, 20, 854
Thomas Culberson, 40, 25, 500, 600, 107
John Wilcox, 25, 15, 200, 20, 75
Charles Pickering, 25, 100, 200, 5, 50
G. Wilcox, 26, 250, 600, 20, 210
Cotton B. O'Neal, 125, 160, 460, 25, 673
Job Freeman, 40, 200, 1000, 25, 760
William Vanzant, 20, 30, 200, 20, 280
Philips Oakes, 20, 30, 250, 8, 90
William Hawkins, 40, 110, 450, 4, 200
Nathaniel Harrington, 22, -, 200, 400, 5, 350
Stephen Harvey, 40, 640, 1000, 20, 1280
Melcher Houston, 20, -, 100, 5, 62
Thomas F. Williams, 130, 110, 3000, 30, 570
Otis Fairbanks, 50, 140, 1000, 10, 270
D. D. Young, 80, 70, 800, 40, 850
Joseph Williams, 100, 300, 2000, 200, 760
M. A. Long, 100, 150, 20000, 250, 1550
R. A. Shrine, 15, 60, 1200, 20, 950
John W. Hale, 250, 483, 4000, 100, 1900
John G. Padricks, 100, 200, 900, 150, 400
James Robinson, 30, -, 10, 5, 150
Joseph L. Branch, 400, 560, 10000, 400, 1530

John D. F. Houcks, 100, 100, 2000, 300, 700
Joseph E. Ritter, 25, 15, 200, 10, 500
Joseph A. Edmondson, 30, 90, 250, 20, 1360
Haley T. Blocker, 35, -, 800, 10, 500
Frederick Toule, 40, 40, 2000, 20, 170
James L. Hart, 220, 180, 2000, 400, 1212
Miller & Brokaw, 80, 80, 1600, 20, 650
Jesse Atkinson, 20, 30, 1000, 5, 110
J. T. Lipscomb, 25, 45, 400, 25, 168
Thomas Laversage, 500, 600, 10000, 740, 250
John Whitehead, 1200, 1105, 22050, 675, 3203
John Cromartie, 250, 550, 4000, 350, 1724
W. L. Stroman, 120, 80, 1000, 15, 290
Frank T. Cromartie, 120, 300, 2500, 20, 525
Bumsby Peacock, 45, -, 500, 20, 230
Wm. Weeds, 20, -, 100, 10, 90
B. Manning, 350, 950, 4500, 700, 1870
G. B. Coleman, 50, 30, 400, 10, 285
William Alderman, 400, 400, 4000, 300, 1750
James Chapman, 17, 143, 400, 10, 200
Edmond Davis, 100, 20, 1000, 20, 1050
Henry Carter, 30, 130, 300, 5, 272
Herring Carter, 100, 320, 2500, 300, 1040
John Grambling, 90, 70, 1600, 825, 600
W. D. Austine, 18, 62, 6500, 5, 250
Henry Bradley, 40, 80, 1200, 10, 335
C. Fletcher, 60, 140, 600, 250, 300
C. Grambling, 100, 160, 1000, 102, 445

Jessee L. Dantzler, 175, 425, 3600, 725, 850

Henry A. Stroman, 100, 220, 3200, 40, 500

John R. Stroman, 70, 130, 600, 20, 283

W. Lafoone, 100, 340, 2500, 120, -

John B. Eliot, 10, 70, 500, 10, 200

Daniel O'Kain, 140, 300, 2200, 200, 732

William Willis, 145, 85, 1000, 115, 400

David Mattox, 90, 70, 1000, 25, 600

Joseph White, 60, 75, 600, 10, 250

Sam. C. White, 160, 500, 2500, 130, 465

Gilford Dawkins, 60, 150, 1050, 20, 455

John J. Wood, 15, -, 150, 25, 228

Hiram S. Hunter, 115, 165, 500, 10, 400

J. J. Mountain, 25, 15, 150, 20, 205

R. Whitaker, 700, 620, 20000, 800, 3000

G. W. Holland, 700, 1000, 14000, 1000, 4000

George Whitfield, 720, 160, 8000, 1200, 3750

Robert Butler, 400, 510, 750000, 3400, 3000

Charles Bennerman, 700, 500, 10000, 525, 2000

Rehesa Oliver, 30, 20, 150, 15, 280

Will N. Saunders, 70, 60, 1600, 10, 450

James M. Cox, 35, 70, 500, 20, 170

Elijah Johnson, 80, 120, 1500, 65, 960

Louis Robinson, 30, 10, 300, 10, 175

Alexander Gady, 30, -, 300, 10, -

Allen Walker, 25, -, 300, 5, -

Benjamin Watkins (Walters), 70, 100, 500, 40, 250

Richard Regan, 100, 300, 2000, 40, 440

W. Dugger, 35, 165, 600, 6, 150

David Redd, 12, 6, 150, 5, 110

W. Freeman, 20, -, 200, 8, 275

James Wilcox, 30, 70, 400, 75, 320

James Cox, 30, -, 200, 5, 175

E. C. Larkins, 20, 200, 600, 10, 227

Henry Colvin, 30, 50, 600, 70, 150

W. Boyd, 11, 9, 100, 5, 75

Joseph Dottey, 10, 190, 600, 5, 100

Cotton Rawls, 30, -, 300, 5,175

John A. Renfroe, 30, -, 200, 36, 900

James A. Brown, 20, 5, 200, 10, 227

Council B. Allen, 100, -, 1000, 60, 850

D. Boatwright, 24, 300, 500, 10, 100

Benj. Hale, 100, 220, 2000, 600, 505

Benj. Byrd, 150, 50, 750, 125, 300

James Hale, 40, 120, 300, 20, 26

Charles Aligood, 25, 25, 300, 10, 210

Noah McElvy, 30,-, 200, 5, 345

Green E. Willis, 20, -, 100, 10, 130

William Fisher, 280, 50, 3000, 360, 700

William D. Moseley, 150, 100, 3000, 100, 200

J. T. J. Wilson, 120, 220, 4000, 50, 250

Geo. C. L. Johnson, 100, 220, 3200, 50, 380

William Bloxhon, 400, 300, 6000, 530, 1955

William Johnson, 400, 1100, 6000, 125, 735

Jabes B. Bull, 356, 65, 1650, 60, 250

James Addison, 80, 160, 1500, 75, 220

Amos Robinson, 30, -, 1500, 145, 200

Kenneth Be_bry, 450, 670, 10000, 500, 1900

Joseph Gordon, 90, 210, 2000, 200, 200

Samuel Fitts, 80, 80, 1500, 20, 300

John P. Coles, 60, 100, 1000, 50, 500

Gabriel Houcks, 35, 45, 300, 5, 150

Wm. Frierson, 50, 30, 500, 10, 250

R.M. Frierson, 200, 280, 2500, 405, 830

N. Scott McGehee, 240, 260, 3000, 75, 400

Green H. Charies, 600, 500, 11000, 430, 2740

Joseph Barrow, 29, -, 200, 30, 400

Benjamin Stevens, 12, -, 100, 4, 60

R. Turner, 15, -, 100, 40, 200

Thomas G. Carmine, 20, -, 300, 25, 400

Eduard Grantham, 15, -, 100, 5, -

Louis Grantham, 15, -, 100, 5, 150

Nicholas Ellis, 30, -, 100, 4, 250

Bethel Grantham, 25, -, 200, 12, 200

Elias Grantham, 15, -, 100, 10, 45

James DeWitt, 37, -, 150, 20, 200

Allen Faircloth, 50, 150, 500, 80, 660

Spencer J. W. Roach, 140, 80, 1000, 50, 800

Benjamin Chaires, 600, 400, 12000, 435, 2300

John M. Hill, 80, 80, 1000, 100, 170

Edward Footman, 400, 120, 5300, 500, 1470

David Robinson, 60, 20, 800, 60, 150

Abner R. Isler, 120, 200, 3200, 300, 620

Julia F. Lorimer, 800, 1000, 4000, 2500, 4280

Wm. M. Taylor, 1100, 1000, 15000, 850, 3600

G. W. Parkhill, 720, 300, 10000, 625, 2850

Richard Croom, 600, 360, 10000, 450, 1666

Theodore Turnbull, 300, 1100, 16600, 600, 1068

H. M. Cason, 150, 290, 2640, 230, 1020

Thomas Reynolds, 259, 221, 4800, 1000, 1425

Flavius A. Byrd, 150, 200, 1000, 50, 575

Jams B. Beard, 50, 51, 1250, 50, 495

Lemuel Jones, 60, 20, 200, 75, 542

John Miller, 970, 1220, 16500, 600, 2700

William Perkins, 350, 370, 7200, 360, 1200

George E. Dennis, 260, 248, 4064, 580, 1310

Noah E. Thompson, 700, 339, 5200, 423, 1325

Robert W. Alston, 305, 500, 7000, 250, 1600

Susan Blake, 425, 455, 9000, 448, 2060

Peter R. Baum, 40, 40, 1000, 200, 665

Edward Carmine, 70, 50, 720, 7, 220

James Holderness, 100, 140, 2000, 100, 325

Nathan Holt, 250, 2890, 2500, 235, 790

Willis Holt, 35, 85, 500, 10, 125

Patrick Smith, 222, 378, 5000, 150, 1000

Burton K. Smith, 100, 500, 4800, 650, 626

Jacob Eliot, 120, 280, 8000, 580, 1125

Green Isler, 30, 50, 1000, 8, 110

Chas. Lee, 90, 30, 1120, 549, 1085

Marcia Theus, 150, 130, 1400, 400, 798

William Thompson, 300, 360, 9000, 360, 1260

William J. Felkell, 30, 50, 300, 45, 135

Berrien Manning, 35. -, 250, 10, 120

Adam Grambling, 75, 165, 2000, 125, 500

Andrew P. Grambling, 60, 100, 800, 125, 300

John Felkell, 80, 253, 2600, 450, 220

John R. Crump, 100, 60, 1200, 280, 180

Robert Williams, 400, -, 4000, 575, 6540

John S. Isler, 55, 105, 1600, 70, 360

Polly Brickle, 40, -, 300, 10, 120

William Hall, 400, 960, 5000, 300, 940

Lyman Smith, 80, 240, 2500, 100, 425

John D. Stroman, 130, 130, 1300, 125, 750

James Felkell, 16, 64, 800, 200, 260

Charles L. Louker, 80, 320, 800, 100, 250

Easter B. Bowles, 60, 180, 2400, 40, 470

James Billingsley, 75, 125, 600, 40, 260

David Alderman, 700, 300, 10000, 350, 1355

William S. Murray, 160, 130, 1000, 125, 1200

W. E. Fisher, 230, 490, 4300, 1000, 1208

Bryan Croom, 780, 1240, 19000, 500, 1310

Jacob L. Felkell, 100, 260, 3600, 250, 270

Wesly Forbes, 250, 270, 4160, 575, 1085

Daniel Sweitzer, 90, 110, 1400, 85, 371

Desdemona Johns, 360, 200, 3000, 445, 1220

Sarah Harley, 70, 10, 800, 50, 550

Joseph F. C. Harley, 80, 400, 2500, 200, 880

James J. Whitehurst, 25, 55, 400, 10, 190

Christopher C. Moore, 130, 130, 1500, 100, 520

Louis Saunders, 150, 710, 8000, 435, 485

John Hutchinson, 25, -, 100, 10, 150

William F. Robinson, 650, 510, 9600, 535, 2100

Joseph A. J. Roney, 70, 15, 1000, 125, 300

John C. Colger, 16, 144, 1000, 95, 200

Joel Jeffcote, 80, 160, 2000, 490, 500

Edward Blacklege, 100, 380, 1600, 50, 320

Joseph Christie, 150, 400, 3000, 100, 598

Theodore W. Brevard, 83, 162, 2700, 80, 260

John D. Branch, 75, 50, 1000, 150, 600

Matthew Turner, 310, 660, 6420, 410, 895

Milly Nelson, 25, 55, 225, 35, 225

Thomas G. Gaskins, 120, 160, 700, 75, 800

Margaret Ham, 120, 140, 3000, 50, 725

Nancy Herring, 150, 90, 2000, 25, 560

John S. Hart, 300, 240, 5000, 560, 1290

Sarah Miller, 60, 300, 2500, 250, 513

Thomas Taylor, 70, 250, 800, 100, 350

James Farmer, 55, 30, 250, 20, 360

James Harriede, 30, 40, 300, 30, 220

Jacob Harriede, 15, -, 50, 5, 80

Simone Partridge, 120, 200, 3200, 15, 760

Andrew Taylor, 15, 45, 5800, 100, 1235

Joshua Montford, 100, 140, 1600, 110, 300

George E. Farmer, 250, 300, 2000, 75, 975

John W. Townsend, 30, 370, 3300, 125, 250

Murdocke Pipkins, 45, 55, 800, 80,380

Tobias Jackson, 75, 85, 500, 125, 375

Jonas Turner, 9, 390, 3000, 100, 1040

George Jones, 841, 700, 19000, 1100, 2690

Fred R. Cotton, 1000, 680, 25000, 750, 3350

Brian Croom 800, 800, 16000, 300, 5750

Braiden Bryan, 100, 140, 1500, 260, 600

Joseph Hale, 100, 140, 1500, 260, 600

John R. Moore, 180, 287, 2500, 50, 750

Jacob Hay, 60, 140, 600, 20, 364

Robert Levy, 40, 80, 1000, 100, 400

Geo. K. Walker, 60, 140, 8000, 200, 621

Anthony Maize, 22, -, 1500, 150, 290

Richard Hayward, 400, 160, 7000, 200, 1500

Richard Hayward, 700, 500, 20000, 200, 3000

Edgar M. Garnett, 500, 140, 10000, 1000, 2000

Edward Houston, 600, 681, 15000, 600, 2000

Henry B. Ware, 235, 485, 5000, 25, 1545

R. W. Williams, 1850, 2750, 46000, 700, 4500

John S. Shepard, 1850, -, 18500, 300, 5000

Jesse Averitt, 1180, 800, 19680, 840, 3235

John Cason, 700, 820, 15200, 400, 2500

James Hunter, 200, 300, 1500, 350, 920

Silas D. Allen, 115, 85, 1500, 25, 565

Jethro Bradley, 100, 140, 1500, 75, 450

B. F. Whitner, 500, 700, 13000, 1000, 3500

G. A. Chaires, 1800, 3600, 32000, 400, 5400

N. L. Thompson, 1600, 1300, 40000, 700, 5000

Henry Copeland, 250, 770, 15000, 200, 7200

Est. of J. W. Cotton, 1150, 500, 6500, 500, 3000

Furner Chaires, 600, 1560, 21000, 400, 3200

Charles Cole, 260, 260, 5000, 200, 1000

Thomas A. Bradford, 1000, 200, 12000, 50, 2500

W. G. Ponder, 500, 1500, 15000, 600, 2000

Wm. Cannon, 100, 140, 600, 50, 310

Francis Eppes, 700, 1220, 28000, 730, 3325

L. B. Brackett, 160, 38, 1000, 300, 1200

R. H. Bradford, 700, 438, 12000, 400, 2300

William Germany, 250, 390, 4000, 250, 1230

John C. Montford, 350, 300, 8600, 205, 1150

John Q. Cromartie, 300, 400, 900, 520, 1500

John A. Craig, 800, 740, 30560, 1200, 1200

Estate of John Wilford, 100, 60, 700, 40, 500

James E. Broome, 310, 350, 3960, 50, 1800

Solomon Sills, 160, 160, 1600, 150, 500

John G. Gamble, 600, 960, 30000, 400, 1600

T. R. Bettons, 600, 340, 10000, 300, 2000

Joseph Chaires, 800, 440, 24800, 1150, 450

William Lester, 1200, 2100, 25000, 1500, 6400

Samuel Bradshaw, 130, -, 300, 10, 60

Wm. M. Maxwell, 400, -, 3200, 250, 1360

J. J. Maxwell, 600, 320, 9000, 750, 1620

John P. Maxwell, 300, 20, 3200, 255, 1098

John S. Maxwell, -, -, -, 145, 1000

George Galphin, 240, 160, 4000, 470, 1810

William H. Burroughs, 325, 315, 9500, 825, 2957

Walter F. Lloyd, 350, 130, 4800, 720, 2200

Jess M. Robertson, 100, 200, 2000, 120, 450

Daniel Johnson, 100, 500, 4000, 100, 450

John P. Billingsley, 100, 220, 1500, 100, 400

Alex Cromartie, 450, 570, 10000, 300, 1770

Calvin Johnson, 150, 20, 2200, 50, 525

Nancy Johnson, 150, 130, 1000, 300, 440

John H. Shehee, 90, 100, 1500, 30, 400

John H. Rhodes, 80, 120, 1000, 10, 450

Isaiah Johnson, 175, 325, 3000, 215, 780

R. K. Call, 600, 866, 22000, 575, 1415

Richard K. Call, 700, 866, 25000, 350, 3705

Richard K. Call, 65, 600, 2500, 20, 2400

John Cook 140, 580, 7200, 2201, 855

William Chester, 100, 220, 800, 100, 800

Peter L. Barne, 30, 30, 200, 5, 204

Geo. Munroe, 130, 200, 2500, 100, 153

Henry Sexton, 70, 90, 350, 15, 500

John Branch, 700, 540, 12500, 1000, 3000

Nicholas N. Branch, 150, 85, 1000, 170, 460

John Whitehead, 90, 500, 12000, 1115, 4030

Isaac Hutto, 30, 130, 200, 35, 210

Israel F. Beard, 200, 200, 1500, 150, 1500

James Hunter, 400, 560, 10000, 800, 2617

James Willis, 25, 60, 400, 15, 200

Thomas Harvey, 180, 60, 1700, 190, 620

Ed. P. Grant, 90, 40, 900, 135, 620

Francis Johnson, 150, 150, 1000, 50, 475

Thomas Baltzell, 110, 80, 2500, 150, 650

William Johnson, 75, 405, 1000, 40, 485

James Forehand, 65, 175, 1000, 30, 530

Lionel Fletcher, 35, 145, 800, 15, 200

James A. Randolph, 175, 385, 5000, 200, 863

Charles Harriede, 60, -, 200, 60, 200

A. G. L. Hodgson, 80, 8, 300, 25, 335

W. L. Ferrill, 16, -, 200, 20, 210

Jesse S. Russell, 35, 115, 500, 55, 208

Martha Chaires, 650, 250, 10000, 400, 2887

George T. Ward, 1450, 990, 25000, 600, 7000

Edward Bradford, 1200, 1800, 30000, 600, 3430

T. B. & C. Chaires, 1400, 1300, 84810, 50, 3700

Henry Bradford, 700, 800, 20000, 1000, 3700

Alman Levy, 50, -, 400, 150, 650

H. Hudson, 30, -, 100, 15, 200

Jacob Chaser, 35, 55, 800, 20, 125

L. H. Branch, 350, 450, 8000, 2000, 3200

Parker Levy, 80, 40, 1000, 100, 730

.

Levy County, Florida
1850 Agricultural Census

The University of North Carolina at Chapel Hill filmed the 1850 agricultural census for Levy County from originals at the Florida State University under a grant from the National Science Foundation in 1963.

Columns 1, 2, 3, 4, 5, and 13 represent the following information on the census:
1. Name of Owner, Agent or Manager of Farm
2. Acres of Improved Land
3. Acres of Unimproved Land
4. Cash Value of the Farm
5. Value of Farming Implements and Machinery
13. Value of Livestock

Isaac Highsmith, 15, -, 150, 100, 350
Jos. Wilkinson, 15, -, 150, 100, 500
Willis R. Madison, 20, 20, 200, 100, 500
Sebastian Tomlinson, 30, 130, 400, 75, 550
John Wester, 15, 25, 200, 50, 500
Aaron Weeks, 25, 15, 200, 50, 240
Wm. J. Hart, 20, 20, 100, 20, 200
Ephraim Morgan, 25, 55, 200, 75, 400
Thomas C. Barrow(Barron), 30, 130, 2500, 250, 1180
Anna Hagan, 15, 25, 200, 75, 400
Nelson McDonald, 10, 30, 200, 50, 100
John Rodgers, 20, 20, 200, 25, 350
Thomas Poor (Love), 70, 250, 2500, 150, 770
Benj. Brownlow, 16, 24, 500, 75, 235
Simeon Harvey, 20, 20, 200, 50, 275
Isaac P. Hardee, 10, 30, 200, 30, 170
Henry Lancaster, 7, 33, 100, 30, 375
Wm. Mayo, 10, 30, 200, 30, 300
Sylvester Bryant, 40, 120, 1200, 200, 1675
Samuel M. Clyett, 125, 1284, 10000, 200, 8840
John Waterson, 40, 120, 800, 75, 1100

Christian Trosper, 10, 30, 200, 10, 400
Lucretia Raulerson, 10, 30, 200, 10, 500
Eliza Hunter, 10, 30, 200, 10, 500
James C. Barco, 20, 20, 200, 10, 300
Emanuel Studsell, 10, 30, 200, 10, 200
Samuel Cowden, 10, 30, 100, 15, 230
Mary V. Andrews, 30, 130, 500, 300, 1300
Enoch Daniel 12, 28, 100, 250, 1300
James G. Daniel, 14, 26, 100, 50, 350
Alfred Mooney, 11, 149, 300, 10, 325
Daniel A. Morgan, 15, 145, 800, 25, 500
Wm. Tomlinson, 20, 200, 200, 50, 1880
John R. Hatcher, 8, 32,100, 25, 125
Moses Cason, 28, 132, 500, 30, 1070
Zachry Davis, 14, 145, 1000, 75, 250
Hy. H. Johnson, 10, 150, 300, 40, 600
James McGeehee, 10, 30, 100, 40, 275
Vincent Johnson, 10, 30, 100, 10, 240
Anthony Ivy, 20, 140, 700, 40, 800

Thompson & Presler, 200, 770, 10000, 1000, 1470

John Goree, 18, 148, 2000, 200, 275

Madison County, Florida
1850 Agricultural Census

The University of North Carolina at Chapel Hill filmed the 1850 agricultural census for Madison County from originals at the Florida State University under a grant from the National Science Foundation in 1963.

Columns 1, 2, 3, 4, 5, and 13 represent the following information on the census:
1. Name of Owner, Agent or Manager of Farm
2. Acres of Improved Land
3. Acres of Unimproved Land
4. Cash Value of the Farm
5. Value of Farming Implements and Machinery
13. Value of Livestock

William L. Took 400, 600, 5000, 600, 2000

Rebecca S. Dozier, 300, 340, 10000, 1000, 1400

John C. McGehee, 600, 1850, 25000, 2000, 3000

Robert Parramore, 120, 80, 500, 100, 400

Daniel Burnett, 40, 40, 400, 60, 900

David Henderson, 23, 17, 310, 50, 400

William P. Mosely, 200, 720, 5000, 300, 1000

William Ramsey, 10, 30, 100, 20, 200

Thomas McDonald, 65, 100, 1000, 100, 600

John H. Patterson, 30, 10, 200, 40, 300

John H. Bryan, 110, 260, 8000, 300, 700

John McLeod, 20, 20, 200, 20, 120

Coleman Roe, 20, 20, 100, 30, 300

Ira Swift, 40, -, 425, 100, 1600

Robert McKinney, 30, 10, 300, 20, 200

William C. Miller, 100, 300, 220, 300, 400

Joshua Loper, 10, 30, 100, 10, 160

John Coker, 43, 7, 200, 40, 200

Hiram Parish, 20, 20, 100, 20, 200

Thomas Langford, 20, 20, 200, 20, 360

John Carmichael, 50, 400, 1200, 60, 500

Francis Agnue, 4, 16, -, -, -

Zechariah Deal, 27, 13, 300, 20, 200

Calvin Loper, 25, 15, 200, 40, 300

William Pridging, 20, 20, 100, 20, 146

James McNiel, 18, 22, 150, 20, 120

James Allen, 40, 40, 2000, 150, 770

John D. McLeod, 120, 840, 6000, 300, 1200

William Henderson, 13, 20, 200, 30, 200

Samuel Hinton, 35, 40, 500, 20, 200

Joshua McCall, 40, 40, 5000, 10, 500

William A. Moseley, 400, 640, 6000, 300, 1000

Hampton Sessions, 30, 10, 200, 30, 200

John Shurrard, 20, 20, 200, 400, 2600

Dennis Hankins, 450, 790, 10000, 1000, 2600

David Patterson, 70, 30, 500, 60, 400

Jeremiah Hinton, 95, 140, 1100, 200, 400

William Gaskins, 23, 17, 200, 20, 200

Alexander Gaskins, 20, 20, 200, 40, 200

William Seaver, 100, 320, 3000, 300, 1000

Nancy Overstreet, 20, 20, 300, 40, 200

William A. Brinson, 210, 300, 1200, 250, 1400

Theodore Hastesily, 500, 700, 12000, 250, 1800

Burton C. Pope, 100, 220, 1500, 400, 1000

Henry E. Ardis, 170, 240, 2000, 300, 760

Charles Begg, 120, 80, 2000, 200, 800

Godlin Perdew, 45, 100,600, 100, 300

William H. Smith, 81, 20, 400, 100, 300

John L. Smith, 35, 5, 300, 20, 160

Lassure L. Fussell, 30, 20, 2500, 160, 300

Andrew Charles, 20, 20, 3000, 60, 1400

Andrew Lannion, 20, 20, 200, 30, 10000

John Sullivan, 30, 10, 200, 40, 400

John Sapp, 60, 20, 500, 200, 2000

David Sestrunk, 100, 40, 700, 100, 600

Joseph Locklier, 40, 40, 400, 40, 300

Paul Hatch, 40, 40, 500, 60, 600

Irene Jones, 30, 30, 200, 50, 400

Arestas Jones, 60, 20, 400, 100, 300

Streit Stampson, 30, 10, 200, 40, 100

Thadious E. Dean, 50, 250, 3000, 200, 4000

Andrew Webb, 50, 40, 350, 75, 350

William R. Hays, 30, 40, 300, 60, 450

Andrew Chambles, 40, 40, 400, 100, 1400

William Biven, 40, -, 500, 30, 800

David Grinto, 50, 30, 300, 60, 451

William P. Lightfoot, 30, 10, 150, 30, 200

Nathaniel Grinto, 70, 20, 180, 40, 300

Dennis Driggers, 20, 20, 100, 30, 180

Simon Driggers, 10, 30, 1000, 20, 100

Eligah Grinto, 30, 10, 200, 40, 900

Denis Driggers Sr., 50, 30, 500, 40, 1100

Manas Williams, 40, -, 300, -, 675

James Burns, 30, 10, 200, 60, 370

Moses Clemon(Cleman), 60, 110, 375, 100, 460

Robert Allen, 30, 10, 200, 31, 220

Zanoc Garner, 50, 30, 600, 60, 200

Israrel Toussant, 45, 35,500, 40, 250

Thomas Burnett, 100, 60, 1000, 200, 1000

Glison Williams, 50, 110, 400, 90, 575

Duncan Bell, 400, -, 300, 25, 225

George W. Conin, 30, -, 100, 8, 50

Andrew J. Lee, 130, 190, 7000, 300, 1000

Thomas A. Shehee, 170, 430, 3000, 260, 1100

John Wilder, 30, 10, 300, 60, 200

Thomas Wilder, 20, 20, 200, 40, 170

Francis Roundtree, 50, 30, 300, 30, 3375

Nathaniel Willard, 10, 30, 4000, 200, 2000

John Toule, 30, 60, 600, 60, 140

Sylas Coker, 20, 20, 200, 30, 200

Thomas Newburn, 20, 20, 150, 60, 600

Giles Estus, 30, 10, 100, 40, 400

Lemon Durrene, 20, 20, 200, 30, 200

James Henderson, 40, 20, 500, 60, 800

James Milleus, 20, 20, 200, 40, 700

John Alking, 75, 225, 700, 170, 300

James Lasley, 20, 20, 300, 140, 200

Richard Croford, 40, 40, 300, 100, 240

William C. Goff, 100, 200, 1200, 140, 500

Anderson Oneeds, 20, 20, 100, 40, 200

Nansey Howard, 40, 40, 300, 63, 240

Mary P. Collins, 80, 80, 400, 150, 600

Richard J. Mays, 700, 230, 15000, 600, 4000

Abram Stephens, 50, 30, 1000, 140, 460

Whitten _. Hine, 170, 90, 3000, 200, 1000

Thomas Chastain, 100, 40, 450, 60, 250

Robert Henderson, 20, 20, 200, 30, 400

Ruley Sapp, 30, 30, 400, 40, 500

James R. Grinto, 25, 15, 200, 30, 200

John Ragan, 30, 30, 600, 150, 500

John Grambling, 40, 130, 500, 150, 475

John A. Q. Collins, 60, 120, 300, 110, 400

James Rambo, 90, 10, 800, 470, 440

Andrew J. Jarris, 30, 10, 200, 20, 300

Samuel Ragan, 20, 20, 300, 40, 300

Willis Brocks, 25, 15, 200, 100, 200

James Davis, 20, 20, 200, 30, 200

John _. Bradly, 400, 1000, 7000, 400, 1200

William Keeklin, 30, 30, 300, 40, 140

Spirus Cone, 20, 20, 200, 20, 200

James Lunday, 20, 20, 300, 20, 164

David Sanders, 60, 190, 1000, 150, 900

Ryding G. Mays, 750, 1050, 15000, 751, 3920

Reddings W. Paramos, 1400, 3000, 35000, 2000, 20000

Namos McLeach, 20, 20, 300, 30, 400

Vann Randall, 260, 440, 5520, 400, 1400

George Gardner, 100, 200, 1200, 100, 800

David L. Henry, 40, 40, 300, 60, 1000

Hyram Winderson, 30, 30, 300, 31, 400

George W. Hix, 24, 18, 200, 20, 300

Isaac Banting, 885, 460, 1100, 800, 3000

John Limpscomb, 1350, 3630, 49000, 4000 6000

James Limpscomb, 467, 780, 10000, 1000, 2000

Benjamin F. Whitner, 100, 500, 3000, 200, 675

Gibson Lanier, 75, 85, 2000, 20, 200

William Willson, 32, 60, 200, 20, 200

Veneto H. Mays, 100, 120, 1000, 100, 600

John L. Miller, 60, 20, 500, 100, 250

Roger McKiney, 45, 15, 200, 10, 125

Nathaniel J. Sanders, 30, 100, 1200, 100, 360

Thomas L. Linton, 690, 2690, 24000, 1000, 7000

Sarah Bobo, 150, 120, 1700, 20, 800

Ann C. Cooper, 100, 140, 2000, 500, 600

William Williams, 20, 20, 200, 30, 120

Nansey Bishop, 20, 20, 400, 30, 200

Jacob Bugg, 40, 20, 300, 60, 200

David Goodman, 20, 20, 200, 70, 240

Thomas Lea, 20, 20, 200, 20, 140

Robert Goodman, 30, 10, 300, 20, 250

Henry Lewis, 75, 425, 2500, 160, 600

Joseph Haton, 40, 40, 300, 100, 400

Dorling Sapp, 30, 10, 300, 20, 160

William Sapp, 20, 20, 200, 30, 140

Hugh Hardin, 30, 10, 200, 40, 200

Ward Poffwell, 20, 18, 300, 30, 300
Jacob Reason, 45, 25, 1000, 125, 360
William Fulford, 20, 20, 400, 46, 200
John Sapp, 40, 20, 30, 20, 200
Jacob Mott, 40, 20, 300, 100, 500
William M. Johnson, 30, 10, 200, 30, 1600
George Wesler, 20, 20,400, 20, 146
Paul Popwell, 20, 20, 200, 20, 140
Charles Martin, 20, 20, 300, 30, 400
Culep Sapp, 30, 10, 300, 40, 300
William Whidden, 30, 10, 300, 30, 330
Jacing Whidden, 20, 20, 30, 30, 300
Stephen Bridger (Bridges), 80, 40, 1800, 100, 500
John R. Sheffield, 60, 40, 400, 30, 340
Isom Taylor, 30, 10, 200, 20, 400
Loveit Williams, 20, 20, 200, 30, 300
Ebenezer Currey, 46, 34, 400, 100, 300
William Murphy, 22, 18, 300, 20, 200
Danal Coker, 80, 40, 2000, 100, 960
Wiley Landford, 33, 17, 500, 40, 500
Silus Cason, 25, 15, 700, 20, 425
Stephen Whitfield, 20, 20, 100, 12, 180
William Crother, 40, 40, 700, 70, 1100
William C. Crolton, 20, 20, 200, 20, 400
James McLoud, 20, 20, 200, 20, 200
Thomas Long, 20, 20, 300, 30, 220
Rolen Williams, 40, 40, 500, 100, 740
Andrew Blure, 25, 15, 200, 40, 450
William Barnett, 30, 10, 200, 30, 200
Antony Brantly, 250, 230, 4000, 700, 1200
Loucious Church, 600, 1600, 12000, 1000, 2600

Columbus Coffee, 160, 500, 2000, 200, 800
Bryan Coffee, 130, 50, 1000, 500, 1000
Andrew _. Coffee, 90, 60, 1000, 100, 600
William E. Thomas, 138, 440, 8000, 250, 1450
John G. Humphry, 208, -, 1000, 310, 800
Samuel F. Frink, 300, 480, 3000, 300, 1000
John S. Walsh, 300, 120, 1000, 500, 1800
Josephus B. Coffee, 100, 220, 1000, 100, 500
Sylvester Hinton, 21, 19, 200, 30, 300
John B. Manly, 28, 19, 200, 30, 300
Henry Dewey, 60, 130, 600, 40, 600
Robert H. Shaffer, 100, 260, 1400, 250, 700
William W. Goodman, 230, 410, 5000, 300, 1000
Amset Hingston (Kingston), 45, 25, 500, 50, 200
Fountain Cone, 30, 40, 600, 20, 300
Samuel Williams, 35, 120, 1300, 130, 460
Edward B. Fox, 150, 70, 2500, 300, 900
Mathew Goza, 100, 60, 1000, 100, 400
Richard Harrison, 550, 750, 10000, 600, 2400
Eligah Quaker, 20, 20, 200, 50, 200
George E. Coleson, 15, 15, 300, 20, 140
Thomas J. Wood, 25, 15, 200, 30, 200
John Wood, 20, 20, 150, 30, 300
Louis Mosely, 250, 550, 4000, 300, 1600
Aleksander Cambell, 20, 20, 200, 20, 200

Bryan Sheffield, 100, 100, 800, 125, 3800
William McEllen, 30, 10, 200, 50, 300
Henry Parker, 20, 20, 200, 30, 300
James Blanton, 20, 20, 300, 30, 200
William Butler, 60, 100, 1000, 100, 400
Stephen White, 20, 20, 200, 60, 600
John Osteen, 30, 30, 500, 70, 600
Green B. Hill, 40, 40, 500, 30, 1000
Joshua Odom, 30, 30, 300, 70, 1100
Jane (James) Wallace, 20, 20, 200, 20, 300
Isaac Parker, 20, 20, 200, 30, 270
Thomas L. Whitlock, 350, 250, 2000, 800, 1400
David Pridging, 80, -, 1000, 600, 700
James L. Wyche, 250, 550, 4000, 600, 1300
Daniel H. Smith, 35, 15, 100, 60, 324
Minza Roe, 20, 20, 130, 20, 660
William Wafus, 80, 160, 1200, 60, 400
Isaac Houston, 80, 80, 1600, 200, 500
Benjamin Lanier, 30, 90, 2000, 100, 700
Archibald Campbell, 35, 45, 350, 20, 100
Samuel Townsend, 25, 2300, 300, 30, 200
Samuel J. Perry, 60, 100, 900, 350, 1400
John Wilder, 50, 30, 500, 60, 400
William H. Hayward, -, -, 550, -, -
Edward Jordan, 40, 40, 400, 60, 400
James Sales (Soles), 120, 80, 600, 100, 600

James Watts, 18, 22, 200, 30, 200
James Livingston, 80, 70, 700, 300, 600
Enoch G. Mays, 400, 1300, 9000, 600, 1600
Thomas O. Kirkpatrick, 20, 20, 200, 60, 400
Benjamin Edwards, 110, 90, 1600, 100, 800
John Broome, 100, 60, 1000, 100, 600
Franklin Bunting, 100, 60, 600, 60, 400
Moses Barker, 60, 60, 1200, 60, 800
Jeremiah Reid, 400, 900, 10000, 600, 1400
William P. Holly, 40, 120, 1000, 100, 800
Allen Roebuck, 60, 80, 600, 100, 800
Anna Monks, 25, 15, 150, 90, 425
Anthony Avant, 30, 20, 200, 30, 200
Solomon Philllips, 40, 20, 160, 400, 200
Mary Warren, 75, 35, 600, 50, 635
John A. Jenkins, 20, 20, 200, 30, 200
Archibald Fair, 300, 420, 6000, 500, 1200
Albert J. Dozier, 130, 270, 4000, 200, 1200
Seaborn O. Sullivans, 190, 530, 10000, 800, 1000
Benjamin Walden, 300, 600, 6000, 500, 2500
John S. Brown, 275, 200, 5000, 350, 1500
John C. Pillares, 200, 320, 2500, 400, 1500
Joseph Flowers, 40, 40, 300, 60, 400
George Paul, 30, 10, 200, 30, 300

Marion County, Florida
1850 Agricultural Census

The University of North Carolina at Chapel Hill filmed the 1850 agricultural census for Marion County from originals at the Florida State University under a grant from the National Science Foundation in 1963.

Columns 1, 2, 3, 4, 5, and 13 represent the following information on the census:
1. Name of Owner, Agent or Manager of Farm
2. Acres of Improved Land
3. Acres of Unimproved Land
4. Cash Value of the Farm
5. Value of Farming Implements and Machinery
13. Value of Livestock

Samuel R. Methere (Mettere), 20, 180, 800, 30, 200

James W. Smith, 20, 140, 1000, 20, 160

Edmund A. Howse, 20, 63, 2000, 50, 1600

Josiah Paine, 16, 144, 500, 10, 450

John Scott, 50, 450, 6000, 200, 500

Blake Harter, 200, 160, 3000, 30, 1400

John G. Reardon, 15, 210, 1500, 50, 175

Wm. S. Mann, 40, 40, 400, 50, 225

James Gamble, 60, 100, 600, 50, 225

Allen Godwin, 6, 154, 800, 20, 400

Robert H. Williams, 65, 95, 500, 100, 900

Francis M. Durance, 25, 135, 800, 50, 1300

Angeline Miller, 40, 120, 1000, 50, 800

Francis W. Durance, 100, 150, 200, 20, 70

James Weeks, 37, 123, 1200, 70, 1500

John Holder, 25, 135, 1600, 40, 125

Abner D. Johnson, 37, 123, 200, 50, 12300

Solomon Moody, 75, 300, 2000, 100, 900

Jesse C. Dupree, 10, 150, 400, 100, 600

Ann M. Hammond, 30, 130, 400, 100, -

Ira H. Harden, 25, 135, 400, 50, 400

David A. McDavid, 25, 295, 2000, 100, 450

John A. Brace, 40, 160, 1500, 100, 600

Wm. C. T. Stephens, 36, 196, 2000, 100, 450

Malcom Galbreath, 20, 140, 1000, 50, 250

Solomon Hale, 20, 140, 500, 50, 400

Jesse C. Sumner, 10, 150, 400, 50, 400

Wm. J. Weills, 15, 145, 250, 50, 600

Drucilla Skipper, 40, 42, 450, 100, 300

Ashley H. Brocks, 80, 200, 2000, 300, 700

John Morrison, 80, 80, 1600, 200, 800

Daniel B. Chapman, 341, 166, 1600, 100, 500

Ephraim Gohannel, 6, 34, 200, 10, 150

Henry W. Dixon, 40, 160, 1500, 50, 800

Winston J. F. Stevens, 35, 140, 400, 20, 150

Neil Furgerson, 40, 540, 350, 100, 520

Wm. H. Durance, 35, -, 350, 10, 400

Ruth Monroe, 35, 125, 1200, 20, 250

Saml. F. Coldwell, 300, 2700, 15000, 950, 1300

John Burton, 200, 280, 3000, 50, 100

James Thompson, 15, -, 300, 40, 380

David Monroe, 30, 130, 800, 20, 300

Norman A. McLeod, 26, 40, 200, 20, 100

Daniel Goin, 120, 320, 2000, 150, 700

John F. Collins, 17, 155, 800, 50, 100

Little B. Branch, 50, 110, 2000, 100, 3400

Silas Bailey, 45, 115, 1000, 50, 950

Seaborn J. Stanley, 75, 1200, 8000, 150, 1500

James E. Ellis, 25, 135, 600, 150, 1200

Jona. K. Stewart, 15, -, 100, 50, 500

Robert W. Stewart, 60, 100, 300, 75, 320

Jackson Tyner, 20, 60, 500, 150, 900

John L. Steward, 40, -, 400, 100, 600

Wm. B. Smith, 10, 150, 800, 50, 300

Wm. Brantley, 45, 115, 1600, 100, 500

Josiah A. Lee, 30, -, 300, 20, 250

Hugh Morrison, 9, -, 100, 50, 600

Roderick Morrison, 14, -, 150, 50, 50

Wm. M. Gahagen, 70, 290, 3600, 100, 500

George F. Cross, 30, 130, 800, 50, 400

Nancy Monroe, 10, -, 100, 20, 16

Peter Hale, 25, 135, 1200, 50, 350

Charles Brown, 125, 955, 9000, 300, 1700

Stephen G. Brown, 35, 125, 800, 100, 500

Thomas Barnes, 50, 150, 2000, 200, 500

James W. Piles, 70, 864, 8000, 100, 500

Francis Barnes(Barne), 16, 144, 1500, 100, 525

John M. Bridges, 5, 315, 2000, 50, 400

Nicholas M. Bradley, 160, 340, 6000, 400, 975

Lemuel Griggs, 13, -, 100, 50, 300

John Philman, 150, -, 100, 25, 200

Wm. S. Delk, 25, 55, 500, 100, 1000

Asa Cothron, 20, -, 100, 25, 200

Wm. W. Piles, 40, 120, 1500, 100, 450

James S. Brinson, 12, -, 100, 50, 100

Edwd. F. Denizen, 45, 115, 500, 100, 550

Lucy Duncan, 20, -, 200, 20, 100

Daniel Wilson, 15, -, 150, 50, 100

Nathaniel Hawthorn, 50, -, 500, 50, 350

Paul McCormick, 180, 220, 4000, 200, 950

George M. Payne, 160, 1159, 8000, 550, 700

Wm. R. Robards, 80, 130, 800, 300, 700

George Houston, 50, 200, 2000, 150, 1600

Jack H. Madison, 800, 200, 10000, 3500, 3750

Mary Johnson, 20, 140, 300, 20, 300

William White, 27, -, 200, 20, 500

Matthew Hawkins, 6, 154, 450, 10, 350

John White, 20, -, 200, 10, 400

Elizabeth Mills, 60, 100, 800, 50, 650

Isaac R. Dyale, 25, 135, 400, 50, 300

David Henness, 40, 120, 800, 100, 650

Charles J. Mexon, 45, 425, 900, 200, 400

John D. Henness, 35, 135, 400, 150, 800

Charles Bennett, 18, -, 200, 50, 100

Jams R. Ferrell, 5, 155, 200, 50, 100
George Helverston, 60, 260, 2000, 200, 550
George Rawls, 20, -, 200, 10, 100
Cotton Rawls, 90, 1470, 5000, 200, 1900
George Bright, 20, 100, 400, 50, 300
William Priest, 40, 220, 1000, 50, 600
Jasper J. Willis, 40, 120, 1000, 100, 1200
Benjamin Grantham, 30, 130, 800, 50, 200
James D. Jones, 25, 135, 800, 50, 50
James Connell, 30, 160, 500, 50, 400
Joseph J. Willis, 15, 25, 200, 50, 350
Granville Beville, 25, 135, 200, 100, 500
Allen B. Robinson, 25, 135, 800, 50, 200
Stephen Blocker, 35, 165, 1200, 100, 800
Emral Priest, 40, 120, 800, 25, 350
Gabriel Priest Sr., 100, 220, 3500, 500, 1300
Granville Priest, 25, 295, 1500, 100, 600
Geo. W. Priest, 20, 180, 1000, 50, 2000
Jones (James) Chislom, 12, 1, 150, 75, 175
Daniel A. Hemmingway, 12, 148, 800, 150, 300
John Curry, 25, 135, 800, 100, 300
Jos. P. Smith, 30, 90, 200, 50, 300
Harrison Cronk, 30, 290, 1200, 50, 400
Francis Hagan, 21, 139, 800, 50, 500
Allen Sturdivant, 25, 135, 800, 25, 400
John McKinstry, 120, 40, 300, 450, 1000
Thomas W. Cooper, 75, 325, 4000, 150, 500
James Garner, 30, 50, 600, 60, 350
Joseph Smith, 8, -, 80, 20, 160

John Grinto, 15, 150, 300, 20, -
William Blair, 15, 145, 300, 20, 200
Daniel Edwards, 60, 200, 2500, 400, 1400
Simeon A. Edwards, 15, 145, 1000, 50, 400
Ephraim Blitch, 30, 50, 400, 125, 600
John Curry Jr., 12, 148, 800, 100, 700
Wm. Curry Sr., 40, 200, 1200, 150, 1500
Ethelbert Hagan, 5, 3, 50, 10, 225
Wm. D. Branch, 50, 270, 2000, 175, 2600
William Williams, 25, 135, 400, 10, 150
Joseph Taylor, 11, 149, 500, 20, 130
Benjamin Brinson, 25, 55, 400, 10, 250
Ann Colding, 30, 130, 2000, 50, 250
William Syms, 40, 240, 2000, 100, 280
Nancy Stokes, 40, 260, 1000, 75, 137
Richd. W. Ridaught, 15, -, 150, 20, 250
David Bruton, 50, 150, 2000, 100, 250
James Noblack, 20, 140, 1000, 20, 20
John Tompkins, 60, 300, 2000, 300, 900
John Blitch, 15, 145, 800, 50, 250
Willis Blitch, 20, 140, 800, 40, 135
Frederick S. Lucais, 30, 170, 100, 150, 500
Frederick Meyer, 60, 200, 800, 150, 600
Thomas E. Williams, 16, 143, 800, 75, 500
Benj. A. Weathers, 9, 141, 800, 75, 250
James N. Badger, 65, 395, 2000, 100, 11000

Wm. C. Goolsby, 60, 180, 2500, 125, 384

John S. Gerkins, 30, 130, 1000, 30, 570

Ozias P. Robinson, 30, 130, 1300, 50, 42

Shadrack Atkinson, 50, 110, 200, 50, 1600

Moses Horn, 35, 165, 1000, 250, 400

Richard Miller, 13, 307, 1000, 10, 600

Joshua L. McGahager, 150, 650, 10000, 500, 1400

Chesly D. Hill, 16, 140, 300, 50, 175

James F. Barnes, 70, 30, 200, 20, 300

Abram Geiger, 60, 180, 2000, 500, 1160

John A. Geiger, 50, 190, 2000, 500, 850

John E. Harrell, 14, 150, 800, 75, 430

Jonathan Tyner, 50, 150, 2000, 400, 980

Wm. Tyner, 20, 140, 800, 100, 350

Jane S. Hall, 40, 240, 2000, 200, 1025

William Connel, 70, 330, 2000, 400, 375

George F. Geiger, 10, 30, 200, 10, 100

John Shields, 10, 30, 300, 10, 120

Charles Nix, 50, 190, 2000, 75, 300

Jos. Higgenbotham, 35, 125, 160, 25, 180

Henry Christie, 11, 4149, 200, 75, 745

Henry J. Johns, 16, 144, 200, 25, 170

George Turner, 14, 146, 200,100, 630

John Woods, 10, 150, 100, 50, 500

John Matchett, 12, 40, 200, 20, 60

Edwd. Boyles, 12, 28, 200, 20, 110

Hamilton Boyles, 8, 32, 200, 20, 120

John Allen, 15, 65, 400, 60, 220

George Bessett, 15, 65, 400, 75, 275

Moses Turner, 25, 55, 400, 75, 1225

Wm. Gillis, 10, 70, 200, 75, 200

Wm. Cothran, 12, 28, 200, 50, 250

Jona. C. Stewart, 15, 25, 200, 15, 515

Emanuel Martin, 42, 40, 400, 75, 1500

Owen W. Simmons, 20, 60, 400, 50, 215

Wm. Turner, 20, 140, 400, 25, 900

Lewis McCoskie, 20, 140, 1000, 100, 220

Nathaniel Ellis, 8, 52, 200, 75, 220

John Gaskie, 20, 140, 1000, 75, 220

Micajah Simmons, 8, 60, 200, 50, 390

James Gough, 60, 100, 1000, 2000, 1410

Henry Bealle, 30, 129, 1000, 75, 295

Cader Clarke, 21, 59, 200, 25, 430

Jona. Clifton, 25, 135, 800, 75, 150

Stephen Wells, 20, 20,300, 50, 160

John M. McIntosh, 70, 290, 3000, 300, 600

Nelson Sturges, 12, 308, 800, 10, 275

James A. Dalles, 25, 15, 200, 15, 300

Isaac Williams, 12, 28, 100, 25, 500

Zachry Weeks, 10, 30, 200, 15, 230

Benj. Brooks, 10, 150, 700, 20, 360

Isaac Newton, 10, 30, 200, 25, 150

Aaron Geiger, 15, 25, 335, 100, 325

David Fryer, 15, 25, 300, 10, 650

John C. Shurhouse, 20, 60, 400, 30, 575

James Blitch, 20, 80, 200, 100, 665

Jesse D. Dykes, 20, 20, 200, 75, 800

Gasper Sistrunk, 70, 90, 400, 150, 1180

Spencer Price, 20, 60, 800, 50, 920

Moncrief Bruton, 35, 45, 300, 120, 500

Frederick Bruton, 35, 45, 300, 800, 510

Isaac Wineguard, 15, 25, 100, 25,200

Allen Osteen, 12, 28, 300, 50, 1070

Alex B. Sanchez, 20, 20, 250, 50, 775
John D. Osteen, 30, 935, 2000, 100, 1100
Hy. A. Wickwire, 25, 15, 250, 75, 1300
Abram J. Osteen, 15, 25, 200, 75, 700
Neil McNeil, 20, 140, 300, 25, 475
Wilie Brooks, 100, 900, 5000, 500, 1700
Lewis C. Gainer, 150, 742, 5000, 100, 500
James Adams, 40, 40, 200, 75, 75
Hugh Y. Neil, 27, 53, 400, 30, 85
Bronson L. Lewis, 30, 50, 1000, 75, 200
Kincheloe Adams, 30, 130, 1500, 75, 400
Charles L. Branch, 10, 30, 200, 25, 250
Wm. M. M. Miller, 15, 25, 150, 25, 125
John E. Croft, 8, 32, 100, 25, 150
Gladney Neil, 20, 100, 600, 100, 345
John Neil, 25, 115, 1000, 25, 125
Stephen Bryan, 300, 940, 15000, 500, 2334
Wm. Cothron, 10, 30, 100, 10, 130
Samuel W. Stewart, 25, 15, 600, 150, 1200
Matthew A. Stewart, 25, 15, 250, 50, 700
Allen O. Munden, 26, 130, 800, 75, 800
Robert Hays, 15, 25, 500, 75, 400
Mary Monroe, 10, 150, 500, 25, 50
John Knight, 12, 28, 300, 50, 90
Henry White, 15, 25, 100, 20, 90
Arthur Lee, 25, 55, 400, 50, 461
Avander Lee, 25, 55, 400, 75, 350
Wm. Carruthers, 8, 150, 300, 20, 300
George M. Condfre, 9, 31, 300, 75, 250
Ebenezer J. Harris, 60, 160, 1600, 100, 275

Wm. S. Murphey, 30, 10, 100, 75, 75
James Crumb, 20, 20, 100, 25, 200
Hansford D. Dykes, 26, 121, 200, 25, 300
Wilis L. Crow, 10, 30, 100, 75, 550
Allen Godwin, 10, 30, 100, 20, 150
Samuel Carruthers, 78, 82, 600, 100, 2000
James Hull, 11, 29, 100, 50, 225
Thomas C. Kettles, 70, 250, 500, 50, 800
William Rainer, 10, 30, 100, 20, 200
Walter Bealle, 25, 135, 1600, 75, 300
Irby Roberts, 80, 120, 200, 300, 1300
Gilbert Reeves, 30, 100, 600, 75, 260
James Carter, 10, 30, 200, 125
Loughlin Galbreath, 15, 225, 1200, 75, 500
Francis M. Piles, 200, 600, 6000, 700, 1775
James C. Ballard, 19, 141, 1000, 75, 300
Lewis Ballard, 14, 146, 1000, 75, 300
John McTerrel, 12, -, 50, 50, 250
Emanuel Crawford, 14, -, 200, 75, 250
John Gardner, 20, 180, 1200, 25, 120
Jos. G. Hale, 20, -, 200, 75, 350
Wm. M. Johns, 21, 139, 200, 25, 360
Seaborn Godwin, 25, -, 200, 25, 1500
Richard Godwin, 60, -, 300, 25, 400
Wm. Holly, 65, 95, 1000, 150, 650
Wm. L. Mobley, 30, -, 300, 200, 1500
Willis D. Sellers, 24, -, 400, 6, 425
Saml. Stanaland, 15, -, 200, 75, 300
Hugh Stanaland, 15, -, 200, 75, 230
John Mobley, 80, 80, 800, 100, 1750
Geo. R. Moody, 10, -, 100, 10,100
Murdock Morrison, 25, -, 200, 50, 300
Henry G. Massey, 25, -, 200, 50, 200

Spencer T. Thomas, 80, 80, 500, 200, 980

K. C. Morrison, 45, -, 300, 100, 1250

B. F. Meyer, 130, 470, 6000, 625, 1000

Harmon Crum, 20, 180, 2000, 75, 700

Wm. H. Reding, 35, 125, 500, 75, 1000

Jos. Hart, 10, 30, 200, 20, 200

John Bates, 30, 130, 800, 70, 800

John Gainey(Gainer), 27, -, 270, 20, 80

John W. Pearson, 70, 90, 20000, 150, 1400

M. Dr. Hasson agt, 100, 380, 3000, 150, 900

J. J. Harell, 30, 10, 200, 75, 350

Monroe County, Florida
1850 Agricultural Census

The University of North Carolina at Chapel Hill filmed the 1850 agricultural census for Monroe County from originals at the Florida State University under a grant from the National Science Foundation in 1963.

Columns 1, 2, 3, 4, 5, and 13 represent the following information on the census:
1. Name of Owner, Agent or Manager of Farm
2. Acres of Improved Land
3. Acres of Unimproved Land
4. Cash Value of the Farm
5. Value of Farming Implements and Machinery
13. Value of Livestock

Henry Weatherford, 3, -, 100, 5, 20
John Roberts, 3, -, 100, 5, 12
John Thompson, 10, -, 1000, 8, -
Philip Santine, 8, -, 200, 5, -
Henry Geiger, 10, 150, 1500, 10, 20
Jno. P. Baldwin, 5, 5, 1500, 20, 500

Nassau County, Florida
1850 Agricultural Census

The University of North Carolina at Chapel Hill filmed the 1850 agricultural census for Nassau County from originals at the Florida State University under a grant from the National Science Foundation in 1963.

Columns 1, 2, 3, 4, 5, and 13 represent the following information on the census:
1. Name of Owner, Agent or Manager of Farm
2. Acres of Improved Land
3. Acres of Unimproved Land
4. Cash Value of the Farm
5. Value of Farming Implements and Machinery
13. Value of Livestock

Reuben Kirkland, 45, 255, 500, 100, 568
Leonard Geiger, 18, -, 60, -, 260
William G. Buford, 30, 270, 200, 75, 605
David Nelson, 30, 20, 300, 40, 540
John Sykes, 20, -, 60,-, 292
William Sykes, 20, -, 75, -, 230
William Harris, 50, 450, 350, 150, 517
Henry Geiger, -, -, -, -, 380
John Nettles, 40, 85, 150, 30, 281
George Stewart, 350, 1150, 1500, 180, 268
George Nettles, 20, -, 150, 15, 460
Abraham Colson, 30, -, 100, 35, 240
Erasmus D. Teacy, 75, 1400, 1000, 250, 800
Thomas Kane, 15, -, 60, -, 15
Alicott Dan, 10,-, 40, -, 224
James Colder, 8, -, 25, -, 56
William Stewart, 12, 150, 200,-, 90
Robert Caldwell, 10, -, 50, -, 229
Samuel Walker, 55, 185, 400, 60, 700
B. G. Barnard, 40, 600, 500, 15, 240
Susannah Saunders, 20, -, 120, 25, 270
Noah Waters, 10, 250, 100, -, 66
Roger Crews, 10, -, 150, 38, 380
Joshua Geiger, 30, 520, 150, 20, 467

Thomas Higginbothan, 50, 250, 125, 50, 562
Jacob Geiger, 50, 430, 500, 150, 1310
Jefferson Massie, -, -, -, -, 600
Joseph King, 6, -, 200, 25, 182
Thomas Haddock, 18,-, 300, 40, 965
William Haddock, -, -, -, -, 161
Benjamin Libby, 10, -, 125, 50, 130
Joseph Haddock Jr., 15, -, 150, 30, 906
Maria King, 25, 975, 2300, 30, 600
John U. Geiger, 25, 615, 1000, 175, 734
Jeremiah Peed agt., 320, -, 600, 175, 70
Cornelius Bessent, 6, 34, 300, 25, 420
Wayne Swearingen, 15, 105, 200, 50, 1030
George Halzendorf, -, -, 200, 50, 47
Samuel Swearingen, 40, 120, 500, 50,680
Joseph Haddock Sr., 40, 210, 300, 50, 891
E. R. Albert, 115, 1475, 11000, 800, 1580
James W. Bandy, 687, 200, 7640, 1100, 600
Phineas Wilds, 30, 180, 300, 100, 300

William McKendric, 30, -, 300, 100, 430

Thomas B. Higginbothan, 50, 150, 300, 75, 1348

William Lane, 3, -, 20, -, 300

John Antone, 10, -, 100, -, 142

William Vanzant, 25, -, 200, 50, 550

Rachael Brown, 15, -, 70, -, 239

Zecharia Haddack, 20, -, 100, 8, 395

John Johnson, 20, -, 100, 30, 500

Levi A. Carter, 10, -, 75, 25, 202

Edward Roe, 20, -, 200, 20, 522

James Geiger, 225, 350, 950, 162, 830

Matilda Norton, 75, 125, 295, 50, 332

Abraham Mott, 75, 25, 350, 30, 212

Philip Goodbread, 30, 120, 500, 30, 16

Souder Goodbread, 26, 125, 1500, 75, 421

William Blount, 13, -, 75, 25, 187

Lewis Blount, 20, -, 75, 25, 380

Richard Greene, 21, -, 200, 50, 534

John Beck, 18, -, 150, 25, 165

Daniel Norton, 12, -, 100, -, 150

William Adams, 18, -, 100, 25, 480

Zylpha Lang, 100, -, 400, 100, 1355

Thomas Drawdy, 120, -, 100, 35, 193

George Moats, 15, -, 100, 50, 240

Margaret Moats, 25, -, 200, 25, 600

Louis Moats, 15, -, 100, 50, 240

Elisha Wilkinson, 25, -, 200, 25, 660

Elbert Swearingen, 15, -, 775, 25, 380

Nathan Norton, 30, 270, 400, 25, 580

Jones (James) Roberts, 40, -, 250, 8, 650

Mary Gwinn agt., 22, -, 300, 35, 370

Rebecca Danford, 12, -, 100, -, 50

Ezekiel Dyass, 24, -, 400, 63, 507

Benjamin Dryden, 15, -, 200, 63, 507

Jackson Pringle, 12, -, 150, 30, 280

Samuel Pringle, 15, -, 100, 30, 225

Thomas Smith, 25, 55, 200, -, 20

Josiah Lewis, 20, -, 250, 75, 500

Hower Brady, 130, -, 200, 50, 390

Noah Allbritton, 30, -, 200, 42, 307

James Sapp, 25, -, 175, 75, 89

Hester Louther, 40, -, 150, 50, 516

Spicer Braddock, 40, -, 200, 120, 1112

Joel Tiser (Tison), 100, -, 500, 120, 890

James Osteen, 45, -, 250, 20, 100

A. J. Higginbothan, 30, -, 350, 100, 485

James Higginbothan, 20, 280, 500, 50, 840

Summer Higginbothan, 10, 290, 200, -, 295

David Higginbothan, 15, -, 200, 50, 273

George Sauls, 20, -, 150, 45, 307

Joseph Hagon, 25, -, 100, -, 220

Jehu Boothe, 35, 40, 300, 140, 820

Edy Pendarvis, 25, 255, 200, 80, 645

Louis Greene, 20, 260, 300, 80, 482

Aldrich Braddock, 40, 280, 350, 125, 2125

John Worley, 15, -, 75, 40, 345

John Jones, 60, 290, 500, 50, 1163

John Higginbothan, 25, -, 300, 50, 864

William Crozier, 50, 790, 300, 50, 995

Cyrus Briggs, 15, 285, 100, 50, 240

James Lord, 40, 460, 150, 100, 430

John S. Braddock, 10, 290, 100, 50, 900

William S. Braddock, 20, -, 200, -, 150

Alexander Braddock, 25, 275, 300, -, 560

David Higginbothan, 20, 330, 200, 60, 536

William G. Braddock, 10, -, 100, 25, 70

Hutto Braddock, 100, 540, 500, 40, 600

Joseph Higginbothan, 40, 60, 300, 100, 1270
Hugh Hyland, 16, -, 50, 60, 274
Nathan Wingate, 25, -, 300, 50, 1000
Martha Braddock, 60,780, 250, 20, 215
Martin Whitemore, 12, -, 60, 40, 60
John D. Braddock, 30, -, 400, 130, 624
William Wingate, 30, -, 300, 60, 1990
John Wingate, 10, -, 150, 30, 540
Phineas Johnson, 8, 292, 100, 25, 110
Wm. Braddock, 135, 480, 350, 100, 612
John Tyson, 12, -, 25, -, 180
Owen Wingate, 30, 470, 100, 91, 645
John Johnson, 30, -, 250, 50, 709
James Wilson, 20, 340, 600, 250, 210
William Vaughn, 12, 30, 50, -, 97
John D. Vaughn, 178, 1022, 1000, 110, 260
William Russel, 30, -, 100, -, 76
Thomas Wingate, 6, -, 100, 50, 355
Elizabeth Wilds, 10, -, 100, 50, 580
Nathaniel Wilds, 40, -, 250, 75, 410

James G. Cooper, 300, 540, 1500, 500, 770
Samuel Clark, 200, 30, 3000, 300, 478
James T. O'Neill, 400, 300, 4000, 400, 1165
Ephraim Harrison, 400, 1000, 3500, 500, 560
Margaret Sterratt, 120, 480, 1500, 50, 650
Robert Harrison, 500, 250, 12000, 1035, 906
Daniel Vaughn, 220, 80, 2000, 150, 500
Domingo Accosta, 120, 580, 2500, 150, 220
John A. Caredo, 65, 95, 1000, 30, 110
John Capo, 30, 270, 500, 50, 320
Francis Pous, 200, 930, 2090, 150, 500
Henry Bacon, 190, 194, 2500, 300, 750
Mary Dubose, 250, 250, 2000, 100, 630
Thomas Dubose, 30, 220, 350, -, -
Harrison Sterratt, 55, -, -, 40, 200
William Cooper, 20, -, 200, 25, 300

Orange County, Florida
1850 Agricultural Census

The University of North Carolina at Chapel Hill filmed the 1850 agricultural census for Orange County from originals at the Florida State University under a grant from the National Science Foundation in 1963.

Columns 1, 2, 3, 4, 5, and 13 represent the following information on the census:
1. Name of Owner, Agent or Manager of Farm
2. Acres of Improved Land
3. Acres of Unimproved Land
4. Cash Value of the Farm
5. Value of Farming Implements and Machinery
13. Value of Livestock

Nicholas Sheppard, 40, 10, 300, 100, 400
James Martin, 20, -, 100, 50, 2400
Daniel J. Thomas, 30, 150, 500, 100, 150
Augustus J. Van, 24, 135, 500, 75, 200
John Hughey, 11, 70, 500, 25, 900
John Simpson, 25, 175, 1000, 120, 850
Henry DeMasters, 10, 150, 400, 35, 400
Vincent Lee, 10, 150, 300, 40, 150
Henry A. Crème, 15, 165, 800, 45, 400
Algiran S. Speer, 40, -, 1000, 200, 800
Aron Jernigan, 35, 1225, 400, 150, 7500
Willoughbee Mastew (Mashew), 30, 130, 300, 75, 1400
Isaac Jernigan, 25, 145, 4000, 60, 1200
Wright Patrick, 18, 152, 300, 35, 700
James O. Duval, 80, 300, 10000, 5000, 1320
Thos. J. Stark, 260, 1345, 6000, 2800, 100
S. H. Clay, 15, -, 200, 35, 250
John J. Marshall, 250, 745, 35000, 5000, 3000
John C. Houston, 25, 135, 2000, 150, 3200

Putnam County, Florida
1850 Agricultural Census

The University of North Carolina at Chapel Hill filmed the 1850 agricultural census for Putnam County from originals at the Florida State University under a grant from the National Science Foundation in 1963.

Columns 1, 2, 3, 4, 5, and 13 represent the following information on the census:
1. Name of Owner, Agent or Manager of Farm
2. Acres of Improved Land
3. Acres of Unimproved Land
4. Cash Value of the Farm
5. Value of Farming Implements and Machinery
13. Value of Livestock

Nathan Norton, 35, 5, 500, 150, 400
Joshia Sykes, 40, -, 400, 100, 2020
Henry Henderson, 8, 22, 1500, 150, 1090
John Register, 40, -, 1000, 120, 1940
John Damp__, 60, 100, 600, 100, 950
Gideon Yelvingston, 18, 5, 1000, 500, 3300
Archibald Cole, 250, 2250, 20000, 4000, 3110
Luke Johnson, 18, 25, 400, 50, 824
Jones Curry, 12, 28, 400, 35, 200
John Osteen, 30, 35, 500, 35,-
William Pinner, 18, 140, 500, 100, 1430
John Kirkland, 16, -, 300, 40, 265
John McRea, 20, -, 250, 35, 215
Daniel Ridaught, 30, -, 250, 25, 260
David Higginbothan, 14, -, 400, 40, 3100
John Platt, 16, -, 350, 45, 680
John Smith, 12, -, 450, 55, 1100
William Thigpin, 11, -, 400, 45, 800

St. Johns County, Florida
1850 Agricultural Census

The University of North Carolina at Chapel Hill filmed the 1850 agricultural census for St. Johns County from originals at the Florida State University under a grant from the National Science Foundation in 1963.

Columns 1, 2, 3, 4, 5, and 13 represent the following information on the census:
1. Name of Owner, Agent or Manager of Farm
2. Acres of Improved Land
3. Acres of Unimproved Land
4. Cash Value of the Farm
5. Value of Farming Implements and Machinery
13. Value of Livestock

Thomas T. Russell, 110, 40, 300, 25, 300
Jas. R. Hanham, 10, 690, 800, 100, 800
Antonia Canova, 6, 800, 1500, 100, 355
Michael Usina, 10, -, 200, 15, 300
Celestine Leonardy, 30, 10, 300, 50, 170
John M. Hanson, 300, 700, 10000, 6000, 1200
Joseph Boyd, 4, -, 100, 50, 150
James Sanchez, 50, 100, 1000, 150, 430
Geo. Genopady, 35, 15, 1000, 60, 60
Esteban Arnas, 50, 3, 1000, 100, 830
Rokey Leonardy, 30, -, 800, 40, 370
Antonia Sabata, 35, 200, 500, 50, 500
Francis Andrew, 14, 76, 1000, 120, 1175
Joseph M. Hernandez, 900, 1300, 48000, 50000, 4645
Abraham Dupont, 300, 700, 10000, 2000, 1200
James Pellicere, 120, 980, 5500, 150, 200
James Simms, 15, -, 300, 75, 415

John Ferrira, 12, 20, 200, 30, 350
Paul Masters, 12, 38, 1000, 75, 480
Masericio Sanchez, 30, 320, 1500, 60, 1480
Francis L. Dancy, 100, 2000, 2000, 300, 1220
Manuel Solano, 15, 34, 800, 40, 300
Bartolo Masters, 40, 320, 2000, 200, 1500
Francis Rogers (Rogere, Rogero), 12, 28, 200, 30, 1200
Matheo Solano, 80, 20, 200, 1000, 2140
Philip Weedman, 50, 100, 1000, 250, 1600
Andreas Pacettyson, 40, 60, 1000, 250, 1800
Cetana Solano, 12, 28, 500, 100, 500
John Ashton, 23, 230, 500, 40, 600
Peter Masters Sr., 12, 26, 560,200, 1330
Peter Masters Jr., 8, 72, 400, 75, 625
David Gray, 12, 13, 300, 50,775
Antonio Alvarez, 30, 280, 15000, 300, 400
Cristobal Bravo, 25, 16, 800, 150, 400

Santa Rosa County, Florida
1850 Agricultural Census

The University of North Carolina at Chapel Hill filmed the 1850 agricultural census for Santa Rosa County from originals at the Florida State University under a grant from the National Science Foundation in 1963.

Columns 1, 2, 3, 4, 5, and 13 represent the following information on the census:
1. Name of Owner, Agent or Manager of Farm
2. Acres of Improved Land
3. Acres of Unimproved Land
4. Cash Value of the Farm
5. Value of Farming Implements and Machinery
13. Value of Livestock

Benjamin Marshall, -, 640, 400, -, 350
Thomas J. Snowden, 35, 125, 600, 25, 250
Jacob Johnson, -, 40, 150, 20, 75
J. F. Cotton, 40, -, 500, -, 20
William Dixon, -, -, 50, 20, 250
E. Whitmere, -, -, 50, 250, 100
Burgess Clark, -, -, 100, 7, 400
James Chestnut, 40, -, 150, 20, 200
John Hardin, 40, -, 50, 200, -
Nimrod Glunk, -, -, -, -, 30
Joseph Jernigan, 40, -, 100, 100, 1000
W. L. Cozzins, -, -, -, -, 300
Samuel Vickers, -, -, -, 20, -
Joseph Owens, 12, 230, 50, 30, 200
William V. Flemming, -, -, -, -, 60
Samuel Sowart, -, -, -, -, 50
Forsyth & Simpson, -, 6000, 1500, 700, -
E. E. Simpson, -, -, -, -, 1601
John Kelker, -, 640, 00, -, -
Isaac Hawkins, 70, 120, 500, 100, 550
George O. Fisher, 2, -, -, -, 100
N. L. Sewage, -, -, -, -, 5
W. W. Harrison, -, 800, 500, -, 575
Joseph Forsyth, 5, -, 100, -, 175
Allen McLean, -, -, -, -, 100
James Butler, 15, 25, 200, 20, 150

B__ Jernigan, 18, 225, 500, 100, 350
Isaac McGeehee, 2, -, 10, 5, -
James R. Lee, 3, -, -, -, 600
George Perry, -, 320, 500, -, -
Micajah Sherley, 20, -, -, 10, -
Jackson Martin, 10, 990, 500, 500, 1500
Estate George Lewis, 10, 80, 500, 5, 50
Loften Cotton, 5, 75, 20, 100, 800
Mills Odom, 12, -, 100, 50, 5
Elijah Fuller, 6, -, 100, 50, 150
Silas Jernigan, 2, 38, 1000, 30, 2000
Wa__ Jernigan, 4, -, 100, -, 430
Mrs. M. Jernigan, 10, 10, 150, -, 200
R. S. Allen, 20, 380, 1500, 100, 400
David Henderson, 25, -, 100, 50, 500
Jacob Whitwell, 10, 15, 1500, 5, 200
Cragler Bachelder & Co., -, 12000, 15000, 2575, 900
Benjamin Jernigan, 35, 40, 500, 100, 1000
R. N. Barrow (Barron), 16, 68, 800, 50, 350
William B. Gaines, 6, 50, 1000, 50, 350
Rufus Milligan, 4, 76, 600, 200, 1000
William J. Keyser, -, 1500, 750, 200, 500
Mrs. Whitmire, 2, 12, 100, 5, 600

R. R. Burts (Butts), 2, 118, 1500, -, 75

John McLelland, 14, 80, 500, 100, 250

Jesey Saville, 5, 35, 250, 100, 450

William Stilwett, 2, -, 100, 5, 50

Michael Elliot, 35, -, 150, 15, 150

L. Falk, 25, 60, 150, 100, 200

George A. Clary, 20, -, 100, 50, 40

Elijah McCurdy, 25, 15, 150, 50, 1150

R. T. McDavid, 15, 700, 200, 110, 800

Wyatt Franklin, 10, -, 100, 25, 200

J. Cobb, 20, 20, 150, 50, 800

Gaylor & Campbell, 40, -, 200, 100, 2550

William Sunday, 15, -, 100, 50, 100

C. J. Deake (Denke), 70, -, 100, 15, 200

John Salter, 10, 40, 500, 250, 20

B. Cobb, 40, -, 150, 25, 1300

Mrs. M. Cobb, 4, -, 150, 10,175

B. Pindon, 10, 70, 300, 50, 350

John A. Gregory, 15, -, 100, 50, 50

William Boon, 3, -, 100, 25, 135

Lemuel Webb, 20, -, 100, 20, 300

A. S. Cobb, 20, 20, 200, 100, 900

A. B. Elliot, 20, -, 110, 10, 200

John W. Dune (Dane, Dunn), 4, -, 120, 50, 150

Joshea Snelgrove, 10, -, 50, 5, 100

W. L. Williams, 40, 160, 1000, 75, 435

Alexander Kennedy, 10, 40, 150, 7, 225

Cavy Jernigan, 30, 10, 300, 25, 900

Enoch Nichols, 14, -, 100, 30, 20

John Wilkinson, 25, -, 200, 50, 850

Gaylor & Campbell, -, 40, 100, 25, 800

Alexander Chestnut, 10, -, 100, 10, -

Mrs. Passer (Paner), 10, 100, 500, 15, 10

Mrs. Allen, 7, 33, 100, 5, 560

Elijah Miller, 30, -, 200, 25, 500

E. Cobb, 3, -, 100, 10, 275

John McArthur, 12, -, 150, 10, 260

Allen McKennon, 7, -, 200, 50, 300

Angus McArthur, 3, -, 125, 10,100

Leonard Nelson, 6, -, 100, 25, 60

Edmond Wiggins, 30, -, 100, 50, 400

Neil McMillan, 10, 70, 1500, 100, 280

Jesse Mims, 8, 800, 3000, 200, 450

Reuben Ard, 5,-, 100, 50, 100

Daniel McCaskill, 20, -, 200, 25, 2550

Alfred Shepherd, 14, 26, 200, 5, 500

James Matheson, 20, -, 150, 150, 15

John Matheson, 20, -, 100, 5, 10

Samuel Nichols, 30, -, 200, 100, 217

Armstrong Smith, 10, 30, 200, 10, 250

Charles Elliott, 12, -, 100, 550, 50

John A. Spears, 20, -, 200, 10, 75

Emory Odem, 15, -, 100, 35, 1100

Michael Penoy, 10, -, 100, 20, 90

Edmond Little, 15, -, 125, 10, 50

Jesse J. Donaldson, 20, -, 110, 5, 500

John Frazer, 10, 5, 200, 10, 100

A. McCullough, 17, -, 175, 5, 700

John Pitts, 15, -, 150, 5, 55

Green Ard, 10, 30, 250, 10, 300

Mrs. Spellman, 10, 30, 200, 10, 500

Mrs. Campbell, 20, -, 150, 20, 300

R. McKennon, 15, -, 125, 10, 400

Joel Malone, 14, -, 110, 5, 200

Robert Manning, 15, -, 150, 15, 70

M. McMillan, 10, -, 130, 10, 300

John H. Kelly, 15, -, 100, 5, 30

Jesse Jernigan, 10, 30, 200, 10, 100

Eldridge Jernigan, 20, 20, 150, 10, 150

Hannon Clark, 25, -, 100, 5, 30

Mrs. Camron, 15, -, 110, 5, 75

Neil Wilkinson, 20, 60, 1200, 100, 300

James McCaskill, 20, -, 150, 10, 350

Benjamin Barron (Barrow), 30, 90, 500, 100, 250

Joseph Cobb, 20, -, 125, 100, 175

John C. Julian, 5, 35, 400, 300, 100
Drury Malone, 18, -, 200, 10, 150

B. W. Thompson, 6, 1715, 1000, 320, 900
Daniel Hart, 85, 45, 100, 500, 1500

Wakulla County, Florida
1850 Agricultural Census

The University of North Carolina at Chapel Hill filmed the 1850 agricultural census for Wakulla County from originals at the Florida State University under a grant from the National Science Foundation in 1963.

Columns 1, 2, 3, 4, 5, and 13 represent the following information on the census:
1. Name of Owner, Agent or Manager of Farm
2. Acres of Improved Land
3. Acres of Unimproved Land
4. Cash Value of the Farm
5. Value of Farming Implements and Machinery
13. Value of Livestock

Thomas Bracken, 15, -, 400, 40, 590
E. Maddon, -, -, -, -, 300
A. C. McBride, 12, 120, 60, 10, 956
William McNaught, -, -, -, -, 30
James Ormand, -, -, -, -, 170
Daniel Ladd, -, 400, -, -, 550
N.P. Bemis, 20, 140, 400, 50, 200
A. Denham, -, -, -, -, 40
John Denham, -, -, -, -, 130
Thomas Campbell, 12,-, 100, -, 290
Clara Bateman, -, -, -, -, 130
Allen Faircloth, 16, -, 100, 10, 750
Drew Vickers, -, -, -, -, 360
Lewis Hall, -, -, -, -, 5, 25
Henry J. King, -, -, -, -, 100
E. B. Hall, -, -, -, -, 345
John D. Coleman, 25, -, 150, 15, 450
D. Keneday, 25, -, 150, 15, 322
N. W. Walker, 100, -, 1000, 50, 1000
John Q. Adams, -, -, -, -, 50
Jane Aligood, 45, 232, 225, -, 400
Abijah Hall, 75, 70, 1000, 165, 1334
William Register, 20, -, 200, 40, 220
James Carter, 100, -, 600, 90, 645
Robert West, 45, -, 275, 75, 740
Silas Aligood, 45, 75, 475, 10, 450
William J. Lewis, 45, -, 100, 10, 57
Paul R. Bevill, 125, -, 500, 75, 340
John A. Basco, 50, -, 300, 60, 365
L. W. Moore, 20, -, 1000, 60, 455
A. M. Ferrell, 60, -, 300, 10, 326

S. B. Ferrell, 30, -, 400, 95, 255
Thomas Bratcher, 10, -, 130, 50, 425
Ansel Ferrell, 20, -, 300, 5, 90
John Lewis, 25, -, 100, 10, 147
James Hammet, 80, -, 800, 75, 347
M. D. Page, 5, -, 150, 30, 100
Jonathan Lewis, 150, -, 1000, 30, 380
S. B. Richardson, 350, 475, 7000, 600, 2020
Eli Stephens 40, -, 50, 30, 80
J. M. Gilchrist, 150, -, 1500, 727, 737
E. C. Walker, 40, -, 450, 100, 278
W. H. Walker, 350, 270, 2500, 500, 94
A. T. Gavin, 200, 330, 150, 180, 970
E. Hale, 8, -, 100, 5, 91
J. J. Wiggins, 65, -, 300, 80, 519
E. W. Wiggins, -, -, -, -, 194
Sarah Wiggins, 30, 50, 500, 20, 524
Jeremiah Gray, 12, -, 100, 5, 48
D. S. McBride, 250, 1030, 5000, 525, 1950
Jason Truluck, 80, -, 300, 10, 40
James Guyon, 10, -, 100, 30, 100
Handy Alford, 23, -, 300, 6, 102
Jehu Gideon, 50, -, 300, 30, 576
Charles Gideon, 16, -, 200, 5, 216
Patrick Gideon, 20, -, 300, 20, 310
Wm. J. Coleman, 28, -, 300, 40, 240

Thomas Jump, 25, -, 100, 30, 64
William Lindsey, 25, -, 350, 36, 227
Thomas Mathas, 50, -, 600, 49, 595
Wiley Jackson, 9, -, 150, 20, 80
Albert Jackson, 12, -, 60, 19, 297
B. N. Scott, 100, -, 600, 300, 439
Thomas W. White, 10, 10, 150, 50, 210
Jesse Coggins, -, -, -, 60, 212
John Myrick, 150, 280, 375, 115, 570
S. A. Braswell, 22, -, 300, 10, 290
W. A. Giles, 35, -, 25, 200, 215
Jno. R. Carter, 40, -, 300, 25, 75
Mathew Raker (Roker), 10, -, 200, 40, 306
William Farr, 10, -, 175, 20, 90
Mary Ann Bratcher, 80, -, 500, 50, 528
A. B. Jernigan, 20, -, 200, 50, 122
J. J. Busick, 17, -, 300, 52, 414
J. R. Andrews, 16, -, 150, 3, 232
Wilson Andrews, 22, -, 125, 30, 120
Jas. W. Smith, 150, 160, 1500, 150, 800
M. Braswell, 139, 415, 2400, 350, 982
Henry Mash, 550, 580, 840, 50, 1275
Irvine Sutton, 5, -, 500, 27, 282
W. B. Sutton, 15, -, 350, 5, 140
Penelope Hart, 30, -, 200, 30, 130
Ada Whitstone, 50, -, 200, 30, 310
E. M. Sasser, 10, -, 75, 5, 192
H. G. McAlester, 50, -, 400, 26, 140
E. W. Jackson, 60, -, 60, 40, 100
J. B. Shores, 50, -, 400, 10, 225
A. P. Tully, 16, -, 200, 31, 265
W. H. Mathers, 75, -, 500, 60, 30

A. W. C. Trice, 200, -, 350, 79, 440
M. Jones, 50, -, 500, 20, 212
S. E. Mathers, 100, -, 600, 75, 476
H. H. Walker, 175, 150, 2000, 500, 9033
A. L. Mash, 17, -, 300, 35, 227
W. Coon, 20, -, 20, 25, 255
R. B. Faras, 65, 95, 200, 30, 350
Thomas Ross, 200, 340, 2000, 175, 1590
Henry Darden, 160, -, 1500, 50, 868
Jonathan Matson, 45, -, 400, 15, 206
J. M. Yeomans, 15, -, 250, 35, 631
John R. Miller, 25, -, 400, 27, 415
Cader Kersey, 40, -, 500, 35, 382
William Standford, 10, -, 200, 5, 90
John Grant, 40, -, 600, 77, 384
N. Robarts, 14, -, 200, 30, 96
R. Robarts, 15, -, 300, 40, 300
W. P. Caussasy, 30, -, 50, 30, 300
B. Red, 27, -, 300, 12, 208
James Harral, 10, -, 400, 10, 276
A. Morrison, 40, -, 700, 30, 360
S. J. Houston, 20, 80, 250, 10, 255
Olivia Byrd, 175, 450, 1750, 400, 1016
J. J. Tucker, 135, -, 700, 25, 340
R. Rames, 30, -, 500, 40, 830
Ives Roddenberry, 35, 35, 500, 45, 490
C. Rouse, 20, -, 200, 10, 447
E. Nunn, 20, 40, 200, 75, 280
J. W. Adams, 200, 100, 1000, 100, 1000
J. Sowell, 60, -, 1000, 200, 514
R. Tucker, 70, 10, 500, 200, 525
M. Posey, 25, -, 200, 30, 240

The University of North Carolina at Chapel Hill filmed the 1850 agricultural census for Walton County from originals at the Florida State University under a grant from the National Science Foundation in 1963.

Columns 1, 2, 3, 4, 5, and 13 represent the following information on the census:
1. Name of Owner, Agent or Manager of Farm
2. Acres of Improved Land
3. Acres of Unimproved Land
4. Cash Value of the Farm
5. Value of Farming Implements and Machinery
13. Value of Livestock

A. Douglass, 60, -, 500, 25, 900
A. Campbell, 40, -, 500, 20, 300
Angus Campbell, 70, 70, 500, 40, 300
Kenneth McCaskill, 40, 40, 500, 200, 980
Alex. Tatum, 22, -, 100, 30, 210
John Bowers, 50, -, 300, 50, 385
Daniel Campbell, 100, 20,400, 60, 800
Hugh McDaniel, 30, -, 100, 10, 110
Joseph Connolley, 16, -, 100, 15, 110
Angus Douglass, 25, 105, 500, 40, 105
Malcolm McLucas, 40, 120, 800, 65, 600
Daniel Ray, 25, 100, 235, 12, 375
John Ray, 20, -, 400, 25, 200
John Perall, 40, -, 160, 20, 275
Peter C. McDaniel, 25, -, 75, 15, 300
Donald Guon, 35, -, 140, 25, 300
James McLean, 70, 160, 1000, 300, 375
William Moore, 18, -, 100, 15, 190
Daniel Neil, 15, -, 90, 10, 275
John Campbell, 35, -, 200, 25, 240
Neil Gillis, 30, -, 100, 2, 175
Daniel Gunn, 40, -, 400, 35, 375
Daniel McDonald, 30, -, 300, 30, 290

Daniel G. McLean, 45, -, 500, 100, 600
Benjamin Kemp, 40, 200, 30, -, 175
William Thompson, 15, -, 100, 10, 450
Peter Timmons (Simmons), 70, -, 200, 30, 450
Finley McCaskill, 20, -, 600, 40, 500
Hugh McLean, 30, -, 300, 30, 500
John Barclay, 30, -, 310, 25, 210
Enos Evans, 30, -, 201, 30, 450
Alexander McCollum, 20, -, 200, 10, 175
William McCollum, 15, 30, 250, 10, 251
Arch. McCollum, 40, -, 400, 25,300
N. Moates, 40, -, 160, 20, 175
Daniel McDonald, 30, -, 400, 30, 180
John McIver, 15, -, 150, 20, 175
Anthony Brownell, 15, 40, 50, 40, 300
Daniel McLeod, 25, 55, 300, 30, 180
Daniel J. Campbell, 40, 120, 400, 40, 200
Daniel B. McLean, 20, 150, 150, 31, 175
John Brownell, 24, -, 60, 10, 110
William Williamson, 20, -, 100, 25, 150

Lauchlin McLean, 20, 60, 400, 20, 245

Redison Blount, 25, -, 100, 30, 115

Duncan Henderson, 20, -, 150, 40, 160

McRea Williams, 25, -, 200, 31, 175

John S. McKinnon, 150, 80, 5100, 30, 3000

Giles Bowers, 10, 81, 100, 15, 300

Daniel S. McLean, 40, 160, 500, 30, 225

Alexander McLeod, 21, 181, 310, 30, 330

Sarah McRea, 50, 81, 300, 35, 300

Alexander Anderson, 20, 20, 251, 30, 400

John McDaniel, 30, 50, 301, 25, 375

Daniel McKinnon, 15, 65, 1000, 45, 400

Laughlin McKinnon, 60, 100, 1500, 40, 225

Christian McLean, 18, -, 150, 21, 230

Peter McDaniel, 30, 50, 400, 30, 360

Norman McLean, 20, -, 200, 35, 220

Alexander McGilvery, 17, -, 150, 30, 150

Angus McDaniel, 40, 40, 400, 45, 750

John Anderson, 31, 370, 1000, 50, 301

George Tervin (Lervin), 40, 440, 1000, 100, 475

John McDonald, 15, 225, 1200, 60, 200

Silas Lee, 20, -, 200, 30, 250

John Tiner, 25, 75, 300, 40, 375

Daniel M. McLean, 20, 80, 400, 25, 320

William McLeod, 27, 20, 400, 25, 350

H. G. Ramsey, 40, 40, 300, 30, 900

Isom Walter, 40, -, 70, 10, 170

Jesse Evans, 50, 170, 500, 35, 650

James Evans, 50, 75, 200, 30, 400

Uriah Kemp, 15, -, 100, 15, 295

William Low, 75, 125, 1000, 45, 300

Feilding Sharp, 100, 640, 1000, 100, 225

Benjamin Sharp, 23, 100, 125, 25, 170

David Cumbie, 12, -, 150, 10, 110

Warren Norris, 18, -, 200, 25, 240

O. H. Griffith, 50, -, 150, 50, 300

David Ward, 25, -, 200, 30, 250

Norman McQuagg, 20, -, 100, 40, 275

Alexander Howell, 20, 30, 200, 31, 150

James Bowers, 15, -, 100, 40, 400

John McCay, 70, -, 300, 30, 200

Nathan Land, 12, -, 70, 12, 220

Rebecca Endfinger, 15, -, 100, 10, 151

George Endfinger, 10, -, 75, 10, 50

James Cockeroff, 20, -, 100, 20, 175

Lyttleton Bark, 35, -, 151, 25, 100

Edward Bedsole, 10, -, 75, 10, 225

Henry Cherry, 20, -, 175, 20, 100

George Blackwell, 15, -, 75, 25, 175

Elijah Ward, 25, -, 110, 30, 550

Christian Campbell, 30, -, 125, 21, 640

James Clary, 15, -, 100, 15, 460

Starks Baker, 21, 40, 300, 30, 320

William Ward, 15, -, 75, 10, 456

Jesse Santifit, 30, 11, 160, 25, 300

Samuel Griffith, 30, -, 175, 30, 200

Nortey (Notley) Morris, 4, -, 25, 10, 125

Dan Wetherford, 18, -, 100, 35, 150

William McWilliams, 31,-, 150, 35, 175

John McWilliams, 18, -, 100, 20, 240

William McWilliams, 20, -, 150, 15, 275

James McWilliams, 25, -, 175, 30, 295

David Gartman, 50, 160, 500, 45, 700

Thos. G. Hart, 8, -, 45, 10, 175

Louis Baggett, 10, 31, 161, 12, 740

Edmon Baggett, 40, 40, 320, 20, 530
Nicholas Baggett, 15, -, 100, 20, 540
Owen Baggett, 8, -, 50, 10, 250
Alexander Baggett, 10, -, 75, 8, 200
James Baggett, 9, -, 65, 10, 275
Wright Gaskins, 50, -, 150, 20, 660
Michael Baggett, 20, -, 100, 15, 190
Absolom Baggett, 8, -, 75, 10, 240
Richmon Barrow, 4, -, 100, 8, 120
Enoch B. George, 40, 40, 250, 25, 450
Reubin Hart, 40, -, 150, 30, 245
Elizabeth Gordon, 15, 65, 240, 12, 300
Norman Morrison, 25, 15,160, 20,100
Warren Acre, 15, -, 100, 10, 200
William Stewart, 20, -, 150, 12, 135
James West, 18, -, 100, 10, 750
Timothy Bell, 20, -, 170, 120, 300
Zeph N. Turner, 16, -, 150, 16, 125
Jonah Stokes, 25, -, 150, 20, 200
Peter Steel, 10, -, 75, 20, 300
Robert Morrison, 12, -, 75, 8, 280
Joseph Steel, 27, -, 150, 20, 300
John Williams, 30, -, 175, 25, 200
Bury Briggs, 20, -, 100, 15, 295
Obediah Thompson, 16, -, 75, 10, 200
Darling Johnson, 40, -, 200, 25, 560
Nancy Cothron, 30, -, 250, 30, 450

John Ghent, 25, -, 175, 15, 200
Louis Miller, 20, -, 100, 10, 175
William Cothran, -, -, -, -, 1400
William C. Cothran, 40, -, 200, 28, 5000
Isaac Welch, 8, -, 100, 12, 1300
Michael Welch, 18, -, 100, 15, 275
Zach Nelson, 30, -, 150, 10, 180
John Stafford, 25, -, 100, 25, 600
Elijah Paggett, 15, -, 150, 25, 200
John Willowly(Willowby), 10, -, 100, 15, 150
Philip Gleeser (Gleeson), 25, -, 150, 10, 90
Irvin Singletary, 30, -, 100, 12, 115
Joel Fowler, 35, -, 200, 26, 425
Alexander McRea, 35, -, 175, 15, 210
Gilbert Rey, 12, -, 100, 10, 160
George Paggett, 25, -, 200, 12, 175
John Izrael, 12, -, 150, 10, 112
Richard Strickland, 25, -, 200, 12, 100
Covington Carmer, 25, -, 150, 15, 40
Newsom Drake, 15, -, 75, 8, 110
Nath. Durham, 18, -, 100, 10, 130
Daniel Neil, 25, -, 75, 10, 100
William Bryant, 20, -, 100, 12, 200
Burtron Butcher (Batcher), 15, -, 75, 10, 110

Washington County, Florida
1850 Agricultural Census

The University of North Carolina at Chapel Hill filmed the 1850 agricultural census for Washington County from originals at the Florida State University under a grant from the National Science Foundation in 1963.

Columns 1, 2, 3, 4, 5, and 13 represent the following information on the census:
1. Name of Owner, Agent or Manager of Farm
2. Acres of Improved Land
3. Acres of Unimproved Land
4. Cash Value of the Farm
5. Value of Farming Implements and Machinery
13. Value of Livestock

Frank Kent, 10, -, 50, 500, 100
Alabama Lacy, 25, 55, 800, 10, 460
Seaborn J. Vann, 14, 26, 300, 10, 127
George Baltzell, 50, 150, 1200, 100, 400
William Newman, 40, 150, 150, 15, 125
Robert Potter, 50, 150, 1000, 20, 425
Allen Riley, 10, -, 100, 5, 285
Thomas Bush, 40, 80, 300, 25, 1408
Robert Crofford, 15, 20, 200, 8, 90
Levi Sapp, 20, 20, 200, 18, 50
Porter D. Everett, 300, 100, 6053, 500, 100
Daniel Henderson, 40, -, 500, 10, 155
John Burk(Bark), 25, -, 200, 15, 120
Thomas Bark, 15, -, 80, 10, 475
Frederick Williams, 10,-, 80, 10,110
Washington Tabor, 20, -, 50, 10, 240
Jacob Snider, 12, -, 80, 30, 130
William Slay, 30, -, 300, 10, 116
Johnathan Morris, 12, 38, 500, 25, 226
Sidney Morris, 10, -, 50, 10, 140
James Swindle, 40, -, 500, 55, 209
Jesse Register, 50, -, 250, 25, 365
William Hoard, 20, -, 300, 6, 30
Mortimer Clark 2, -, 50, 5,70
Taylor Bennett, 15, 25, 150, 4, 108

D. W. Horn, 700, 260, 6000, 275, 3850
John Kent, 8, -, 50, 6, 70
Wiley Everett, 250, 400, 5000, 200, 1100
James Dennard, 22, -, 100, 25, 200
Silas Taylor, 20, -, 200, 50, 180
Johnathan Taylor, 30, -, 250, 50, 200
Alex. Brooks, 25, -, 200, 30, 250
William Polten, 40, -, 300, 10, 100
John Burk, 20, -, 80, 5, 95
John Owens, 40, 12, 250, 10, 305
Thomas Owens, 11, 75, 400, 10, 306
Joseph Melvin, 25, -, 80, 15, 100
Joshia Spencer, 35, -, 50, 30, 150
Daniel Melvin, 16, -, 100, 7,100
Jack. Barfield, 15, -, 125, 5, 145
George Gilbert, 12, 28, 100, 20, 95
William Gilbert, 20, 20, 250, 5, 100
Absolem Posey, 42, 38, 1000, 100, 335
Henry H. Wells, 15, 105, 60, 75, 130
John McKithen, 18, -, 1000, 300, 305
Enoch Holland, 23, -, 150, 20, 108
Ennis Pippen, 14, 70, 1000, 25, 450
Samuel Mitchell, 26, -, 500, 50, 320
Solomon Pippen, 23, -, 150, 40, 500
Ezekiel Cooper, 20, -, 150, 20, 110
Amelia Godwin, 150, 150, 2000, 450, 2200

Griffin Taylor, 35, 5, 400, 70, 300
Everett Hill, 12, -, 150, 5, 50
James Swails, 30, 20, 1000, 50, 130
Josiah Jones, 80, 80, 2500, 50, 375
Ashly Vippers (Pippen), 20, -, 300, 15, 136
Solomon B. Evans, 15, 80, 300, 8, 100
James Holt, 60, 40, 200, 55, 230
John McCormac, 30, 150, 1000, 12, 140
P. Parker, 25, 100, 500, 110, 575
David J. Turvin, 17, -, 315, 15, 10
James Brown, 45, -, 200, 45, 2300
Charles Porter, 25, 135, 500, 75, 691
Sharpless Evans, 30, 10, 500, 25, 30
Angus McQuagg, 12, 18, 200, 8, 140
Samuel Gainer, 30, 90, 1000, 50, 353
Casey Taylor, 20, -, 150, 10, 135
George Taylor, 35, -, 200, 15, 408
John Ferguson, 100, 140, 1000, 56, 1100
Stephen Daniel, 60, -, 300, 50, 425
William Ferguson, 30, 10, 500, 15, 250
Teakle Taylor, 41, -, 300, 60, 365
Willis Taylor, 20, -, 200, 10,100
Mathew Taylor, 10, -, 150, 7, 110
Irvin Allen, 10, -, 120, 25, 145
Lenard Tinch, 12, -, 80, 10, 230
Nathaniel Miller, 35, 280, 1015, 40, 485
Jacob White, 90, 200, 800, 25, 345
Isom Pane, 25, 15, 300, 31, 250
Lawson Daniel, 31, -, 250, 35, 190
Martin Worthington, 15, -, 150, 10,105
John White, 20, 380, 1600, 10, 235
James G. Lucas, 60, -, 300, 75, 265
James Hamilton, 250, 130, 2000, 400, 1500
James Long, 180, 440, 180, 200, 1400
Robert Lawrence, 25, -, 430, 35, 800
Thomas Russ, 150, 50, 1200, 75, 885

Druey Braving, 25, -, 75, 8, 190
Alexander Campbell, 25, -, 300, 25, 110
Daniel Brown, 12, -, 150, 8, 160
Jasper Smith, 14, -, 100, 12, 75
Willis Wright, 15, -, 50, 10, 85
John Cook, 60, -, 1000, 30, 420
Thomas Smith, 25, -, 100, 7, 112
Henry Oneal, 80, 145, 1000, 100, 1045
Henry Alford, 20, -, 100, 7, 90
Daniel Smith, 30, 10, 100, 35, 205
Rhoda Register, 5, -, 150, 5, 475
John Register, 30, 50, 100, 20, 320
Gabriel Skipper, 46, 34, 300, 30, 435
Norman Campbell, 30, 50, 600, 75, 465
John Seffield, 20, 100, 250, 100, 935
Young Gilbreath, 20, -, 50, 35, 154
Mary Wise, 20, 60, 200, 5, 510
William Miller, 60, 220, 1000, 50, 671
Robert Minty, 15, -, 100, 16, 130
Levi Miller, 50, 35, 600, 100, 190
Levi F. Miller, 75, 85, 700, 75, 745
Ashly Miller, 75, 80, 1500, 100, 516
John Matthias, 60, 140, 950, 60, 315
Mariah Lockey, 50, 210, 1000, 50, 286
David Palmer, 10, -, 75, 5, 90
William Izrael, 18, -, 100, 7, 190
Sharpless Evans, 60, 20, 500, 70, 385
William Levins, 15, -, 60, 7, 150
Aron Hudson, 20, -, 75, 10, 360
Wiley Godwin, 20, -, 100, 56, 175
Robert Hewett, 35, -, 100, 30, 290
Aron Madox, 35, -, 250, 25, 700
George Hill, 20, -, 150, 20, 140
William Jones, 25, -, 75, 20, 145
Samuel Russ, 50, -, 140, 150, 1150
William McMullen, 15, -, 50, 5, 80
Mary Carr, 5, -, 75, -, 80
Stephen Roche, 35, 365, 365, 40, 1978

John Wright, 30, -, 150, 50, 200
George Wright, 30, -, 100, 20,175
Robert Givens, 22, 5, 100, 4, 150
Brown Lassiter, 35, 65, 300, 20, 600
Brinkly Lassiter, 50, 40, 100, 15, 400
William Winslow, 30, -, 100, 110, 275
Henry Wood, 7, -, 25, 45, 105
John Miller, 30, 57, 200, 30, 535
Samuel Miller, 20, -, 150, 7, 280
Murdock Morrison, 26, -, 200, 10, 112
Shine Williams, 35, -, 100, 35, 575
Iziah Levins, 40, -, 100, 10, 160
Reese Hudson, 20, -, 75, 4,110
Richard Levins, 43, -, 75, 30, 330

Anthony Burns, 18, -, 150, 20,100
Michael Givens, 15, -, 75, 10, 150
Uriah Odom, 16, -, 150, 8, 175
Jerry May, 40, 40, 200, 12, 300
John Tenney, 25, -, 30, 4, 20
Daniel Parker, 30, -, 75, 8, 103
Henry King, 40, -, 100, 12, 135
Charles Bursed, 15, -, 175, 10, 85
Joel Cooly, 18, -, 75, 10, 110
Joshua Mercer, 25, 55, 300, 20, 450
Elizabeth Wells, 15, -, 100, 10, 110
David Sanders, 10, -, 75, 10, 100
David May, 30, 10, 200, 40, 450
Bethel Madox, 20, -, 100, 25, 250

Ruskin, 38
Russ, 46, 49, 94
Russel, 81
Russell, 2, 18, 50, 63, 84
Russer, 41
Rutledge, 17
Ryals, 21
Sabata, 84
Sadbery, 21
Sadbury, 30
Saddler, 22, 35
Sales, 71
Saltenstall, 51
Salter, 86
Sammis, 22
Sanchez, 3, 76, 84
Sanderlin, 12
Sanders, 18-19, 37, 69, 95
Sane, 32
Sanson, 8
Santifit, 91
Santine, 78
Sapp, 3, 11, 13, 68-70, 80, 93
Sarey, 49
Sasser, 38, 89
Sauls, 50, 80
Saunders, 4, 11, 20, 29, 50, 59, 61, 79
Saville, 86
Sawyer, 17
Scarborough, 10, 13, 32
Schaffler, 19
Schornhurst, 19
Scoggins, 53
Scott, 8, 16, 20, 35, 49, 52, 54, 56-57, 72, 89
Scruggs, 52-53
Scurlock, 45
Seaberry, 29
Seabrook, 30
Seals, 6
Sealy, 55
Seaton, 4
Seaver, 55, 68
Secenger, 3
Secrest, 4

Sedgwick, 20
Seely, 33
Seffield, 94
Sellers, 45, 48, 76
Selph, 6
Semmes, 25, 67
Sestrunk, 68
Sewage, 8
Sexton, 63
Shackleford, 53, 56
Shaffer, 70
Sharp, 3, 91
Sharpe, 38
Shaw, 27-28, 36
Shedo, 34
Sheffield, 17, 36, 70-71
Shehee, 63, 68
Shelfer, 34
Shepard, 24-25, 54, 62
Shepherd, 5, 27-28, 30, 50, 86
Sheppard, 31, 82
Sherhoull, 1
Sherhouse, 4
Sherley, 85
Sherrod, 28
Shields, 75
Shirley, 12
Shiver, 47
Shores, 89
Short, 30
Shrine, 58
Shulhouse, 4
Shurhouse, 75
Shurrard, 67
Sibley, 28
Sikes, 15
Silcox, 19-20, 22
Sills, 62
Simkins, 51
Simmons, 6, 8, 26, 75, 90
Simms, 84
Simons, 11
Simpson, 82, 85
Sims, 48, 50
Sinclair, 25
Singletary, 92

114

www.ingramcontent.com/pod-product-compliance
Lightning Source LLC
Chambersburg PA
CBHW082359270326
41935CB00013B/1682